# Applications of Nanobiotechnology

*Edited by Margarita Stoytcheva and Roumen Zlatev*

Published in London, United Kingdom

IntechOpen

*Supporting open minds since 2005*

Applications of Nanobiotechnology
http://dx.doi.org/10.5772/intechopen.82976
Edited by Margarita Stoytcheva and Roumen Zlatev

Contributors
Hamid Reza Rahimi, Mohsen Doostmohammadi, Latih Habeeb, Hasan Sh. Majdi, Lucie Bacakova, Marketa Zikmundova, Julia Pajorova, Antonin Broz, Elena Filova, Andreu Blanquer, Roman Matejka, Jana Stepanovska, Petr Mikes, Vera Jencova, Eva Kuzelova Kostakova, Alla Sinica, Jyoti Sharma, Alaa Yousef Ghidan, Tawfiq Mohammad Al Antary, Yuchao Li, Jiahao Yan, Yao Zhang, Baojun Li, Deepak Kumar Dash, Rajnikant Panik, Anil Kumar Sahu, Vaibhav Tripathi, Baojun Li, Yao Zhang, Urvashi Gupta



First published in London, United Kingdom, 2020 by IntechOpen
IntechOpen is the global imprint of INTECHOPEN LIMITED, registered in England and Wales, registration number: 11086078, 7th floor, 10 Lower Thames Street, London, EC3R 6AF, United Kingdom
Printed in Croatia

British Library Cataloguing-in-Publication Data
A catalogue record for this book is available from the British Library

Additional hard and PDF copies can be obtained from orders@intechopen.com

Applications of Nanobiotechnology
Edited by Margarita Stoytcheva and Roumen Zlatev
p. cm.
Print ISBN 978-1-78985-977-5
Online ISBN 978-1-78985-978-2
eBook (PDF) ISBN 978-1-83880-997-3

For EU product safety concerns:
IN TECH d.o.o., Prolaz Marije Krucifikse Kozulić 3, 51000 Rijeka, Croatia, info@intechopen.com or visit our website at intechopen.com.

# Meet the editors

Prof. Margarita Stoytcheva graduated from the University of Chemical Technologies and Metallurgy at Sofia, Bulgaria, as a chemical engineer and master of electrochemical technologies. She has PhD and DS. degrees in Chemistry and Technical Sciences. She has been a researcher and teacher in several universities in Bulgaria, Algeria, and France. From 2006 to the present, she has participated in activities of scientific research, technological development, and teaching in Mexico at the Autonomous University of Baja California at Mexicali as a full-time researcher. Since 2008 she has been a member of the National System of Research of Mexico, and since 2011 she has been a regular member of the Mexican Academy of Sciences. Her interests and area of research are analytical chemistry and biotechnology.

Dr. Roumen Zlatev is a full-time researcher at the Engineering Institute of the Autonomous University of Baja California (UABC) at Mexicali, Mexico. He obtained his bachelor's and master's degrees from the University of Chemical Technology and Metallurgy of Sofia, Bulgaria, and his PhD degree from the National Polytechnic University of Grenoble, France. He was a full-time researcher at the Bulgarian Academy of Sciences and a part-time professor at Sofia University. In 2005 he accepted the position of full-time senior researcher at UABC. Dr. Zlatev is a member of the Mexican National System of Researchers and a regular member of the Mexican Academy of Sciences. He participates in research projects in France, Germany, and Mexico. Dr. Zlatev is the author of more than 170 publications, book chapters, and reports in scientific congresses and holds 14 patents in the field of electrochemical and spectroscopic methods of analysis, corrosion and materials, and electrochemical and analytical instrumentation.

# Contents

# Preface

This book is dedicated to the applications of nanobiotechnology, i.e. the way that nanotechnology is used to create devices to study biological systems and phenomena. It includes seven chapters, organized in two sections: (1) Nanomaterials in medicine, agriculture and biosensorics, and (2) Nanofluidics.

The first section (Chapters 1–5) covers a large spectrum of issues associated with nanoparticle synthesis, nanoparticle toxicity, and the role of nanotechnology in drug delivery, tissue engineering, agriculture, and biosensing.

Chapter 1 provides an overview on the physical and chemical methods for nanoparticle production with a special emphasis on green technology as an eco-friendly alternative to conventional techniques. It discusses the potential applications of nanoparticles in biomedical fields and attempts to bridge the gap of knowledge on the toxicological effects of nanomaterials.

Chapter 2 comments predominantly on the use of nanomaterials in drug delivery, due to their unique properties, size, and shape, as well as on the great potential of nanotechnology in molecular diagnostics.

Chapter 3 summarizes recent progress achieved in the field of nanofiber-based skin tissue engineering, including results of the author's research group. The review is focused on the properties and use of synthetic non-degradable and degradable polymers and their composites with natural polymers for nanofibrous mesh preparation, intended as scaffolds for skin tissue engineering, wound healing, and carriers for various bioactive molecules.

Chapter 4 addresses two principal topics associated with agricultural nanobiotechnology: The application of nanomaterials to increase agricultural production using nanopesticides and nanofertilizers, and the application of nanomaterials in food packaging. The authors recognize the novel possibilities that the nanotechnological approach offers in agriculture, but also comment on some related constraints.

Chapter 5 introduces two types of optoelectronic devices based on Mie resonances in silicon nanoantennas and analyzes their application prospect in biosensing.

The second section (Chapters 6 and 7) is devoted to the properties of nanofluids and the medical and biological applications of computational fluid dynamics modeling.

The book offers a professional look at the recent achievements and trends in nanobiotechnology applications. All the contributing authors are gratefully acknowledged for their efforts in preparing the book chapters and for their interest in the present project.

**Margarita Stoytcheva and Roumen Zlatev**
Universidad Autónoma de Baja California,
Mexicali, México

## Section 1

# Nanomaterials in Medicine, Agriculture and Biosensorics

Chapter 1

# Nanoparticle Synthesis, Applications, and Toxicity

*Hamid-Reza Rahimi and Mohsen Doostmohammadi*

## Abstract

Nowadays production of different nanoparticles (NPs) with plausible biomedical benefits is tremendously increasing. NPs are of great interest in drug delivery systems, drug formulation, medical diagnostic, and biosensor production. Aside from the importance of NPs in medicine, their negative side effects including potential cytotoxicity, inflammatory response induction, and drug interruption should be carefully considered. Several molecular and physicochemical mechanisms are involved in toxicity induction of NPs. Finding the negative effects of NPs on human tissues and investigation of their mechanism of action are a way for preventing the happening of unpleasant event. Here in this work, we would describe the main way of NP production with special attention to green NP production, and then their application in medical diagnosis and disease treatment would be explored. Also the main toxicity effects of NPs on different tissues would be explored, and the parameters affecting the quality of NPs and their corresponding biological properties would be highlighted.

**Keywords:** nanoparticle, biomedical, drug delivery, cytotoxicity, medical diagnostic

## 1. Introduction

Nanotechnology is referring to the technology of production, characterization, and application of materials in nanoscale [1]. After the definition of this term by Norio [2], nanoparticle (NP) production and application in several different fields gain much attention. The small size and high surface area of NPs are the causes of their tunable physicochemical properties such as improved thermal conductivity, light absorbance, significant chemical stability, and high catalytic activity [3]. Furthermore the surface layer of NPs can be functionalized using chemical and biological agents like small molecules, surfactants, and polymers for enhancing their activity and specificity [4]. It is revealed that each NP, regarding its size, surface charge, shape, surface groups, and type of ions, shows unique biological and physicochemical properties [5, 6]. Owing to such diversity, these materials got immense biomedical applications such as drug delivery, radionuclide therapy, biosensors, cancer therapy, diagnostics, magnetic resonance imaging (MRI), and biological molecules purification [7, 8]. Although a wide variety of NPs with diverse ions and surface modifications are produced and preclinically tested, only a limited number of them gain approval for clinical uses. The long-term stability, general cytotoxicity and inflammatory response induction, and lack of guideline for relevant biological testing are the main reasons for low-approved NPs [6]. Due to the fast development of NPs, it is necessary to identify correlation between the physical and chemical attributes of NPs and their corresponding biological effects. For instance, it is shown that although the positive charge of NPs enhances the efficacy of gene delivery, and imaging, it also enhances the cytotoxicity of corresponding constructs [9]. In this chapter we briefly introduce the main way of NP production and their applications in biological and medical studies. Also the mechanism of cytotoxicity induction and the main ways of detecting this toxicity are explained.

## 2. NP production methods

### 2.1 Physicochemical methods

NPs can be produced through two main approaches including top-down approach which is the production of NPs by making smaller and smaller structures by etching the bulk agents and bottom-up approaches which is the building up of NPs from atoms [6]. Physical, chemical, biological, and in some cases hybrid technique are the main ways of NP production. The physical methods of NP production include methods like laser ablation, high-energy irradiation, spray pyrolysis, and ion implantation, and the chemical one includes chemical reduction technique, sonochemical method, solgel process, microemulsion method, and electrochemistry. The biological method which is also called green NP biosynthesis involves application of plants extracts, microorganisms, enzymes, and even some agricultural wastes for NP production. Although the physical and chemical methods resulted in bulk amount of NPs a few times, application of chemical agents during the NP production in coordination with production of environmentally dangerous compounds simultaneously with NP production limited their applications [10]. For instance, thermolysis which is a chemical method for dissociation of organometallic precursors is performed at high temperatures by using organic solvents. Also in some cases, surfactant is added to the reaction medium for reducing coalescence of particles [11]. Chemical reduction technique is an adopted chemical method which used a wide range of reducing agents such as sodium borohydride, hydroxylamine, and N,N-dimethylformamide for production of zerovalent ions. Wave-assisted chemical method used ultrasonic waves in coordination with surfactant or reducing agent for production of NPs. The formation of micro cavities with high temperatures upon ultrasonic induction can start chemical reduction of substrates. The physical NP production methods are mainly energy intensive and need special devices. For instance, milling process is a way by which metallic microparticles are crush using high-energy ball mills. The gas-phase process or aerosol process which is divided into four main types (including flame reactor, plasma reactor, laser reactor and hot wall reactor, and chemical gas-phase deposition) is a particular way for the production of NPs like fullerenes and carbon nanotubes. All types of these methods need special devices and are mainly high energy consuming. NP production by wet chemical synthesis takes place at low temperature and is one of the most employed methods for NP production. Limitations in increasing batch reactor because of limited mixing and low heat transfer are mentioned as the main disadvantage of wet chemical synthesis method. The main advantages and disadvantages of NP production by physicochemical methods are summarized in **Table 1**.

### 2.2 Green NP production methods

Green technology using biological systems like plants, microorganisms, and enzymes is rising fast as an alternative method for conventional chemical and

| Method | Advantages | Disadvantages | Ref. |
|---|---|---|---|
| Chemical vapor deposition and chemical vapor condensation (CVD and CVC) | • High pure NPs production | • High temperature of procedure (above 300°C)<br>• Uses of chemical agents | [12] |
| Gas condensation | • Production of ultrafine nanocrystalline metals and alloys | • Need for special devices<br>• Extremely slow | [13] |
| Laser ablation | • High-purity NP production | • Need special devices<br>• Difficult to control size, agglomeration, and crystal structures | [14] |
| Solgel | • Simple method<br>• Production of large range of materials<br>• Uses low temperature | • Using chemical agents<br>• Undesirable agents production<br>• The cost of materials may be high | [15] |
| Chemical reduction | • Cost-effective<br>• Good production rate | • Application of toxic agents<br>• Hazardous by product formation | [11] |

**Table 1.**
*The main physicochemical methods of NP production and their corresponding advantages and disadvantages.*

physical. In contrast to physiochemical methods which mainly lead to environmental toxicity, the biological NP production methods are known as eco-friendly and nontoxic protocols [16]. The biologically produced NPs' special features including high catalytic activity, low toxicity contaminations, high stability, and plausible biocompatibility and biodegradability make them distinctive from NPs produced from other methods. The microorganism's related NP productions are classified into intracellular and extracellular synthesis methods [17]. In an intracellular way, ions of interest are transported into the microbial cell and then reduced in the presence of enzymatic processes, while the metal ions are entrapped and reduced at microorganism's surface in an extracellular way [18]. Microbial NP production regarding the ability of the majority of bacteria and fungi in tolerating ambient conditions of varying temperatures, pH, salt concentrations, and pressures makes this approach a safe, cost-effective, and environmental method. Several microbial species have been isolated from different environments and used for production of various NPs. Compared to microbial production method, plant NP production is more desirable because it does not need any special and multistep processes, it has faster production rate, and it has easy scaling up procedure and because of its cost-effectiveness [19]. Investigations have revealed that metals bioaccumulated in plants which sometimes are called phytomining are mainly in the form of NPs. For instance, high level of silver NP accumulation in *Brassica juncea* and *Medicago sativa* [20], gold NP production in *M. sativa* [21], and copper NP accumulation in *Iris pseudacorus* [22] has been reported. This type of NP production has several disadvantages including heterologous size and morphology of NPs, difficult extraction and isolation procedure, and low production yield [21]. The alternative approach is in vitro production method which is based on reduction of ions using plant extracts. This method is more controllable through making change in plant extract and ion concentration, time of reaction, temperature, and pH of reaction medium. The production rate of this method is much faster and easier than in vivo method [23, 24]. For example, the extract of *Tectona grandis* seeds was used for reduction of AgNO3 to 10–30 nm Ag NPs with significant antibacterial properties [25], whereas Au NPs with an average size of about 3 nm have been synthesized using leaf extract of *Ziziphus zizyphus* [26]. Various plant extracts have been used for production of NPs from different ions with diverse sizes and shapes [27]. **Table 2** summarizes some examples of NP production through a biological way.

| Biological entity | Type of NPs | Size (nm) and shape | Special characteristics | Ref. |
|---|---|---|---|---|
| Bacteria and fungi | | | | |
| *Delftia sp*. SFG | Bi | Sphere/40–120 | Antibiofilm activity against *P. aeruginosa* | [28] |
| *Escherichia coli* | CdS | Spherical/2–5 | | [29] |
| *Botryococcus braunii* | Cu, Ag | Sphere/10–100 | Antibacterial and antifungal effects against *Pseudomonas aeruginosa (MTCC 441), Escherichia coli (MTCC 442), Klebsiella pneumoniae (MTCC 109) and Staphylococcus aureus (MTCC 96), Fusarium oxysporum* | [30] |
| Yeast strain MKY3 | Ag | Hexagonal/2–5 | | [31] |
| *Fusarium oxysporum* | Ag | 25–50 | | [32] |
| *Bacillus mojavensis* | Ag | 105 | High antibacterial activity against multidrug resistant pathogens | [33] |
| *Aspergillus fumigatus BTCB10* | Ag | 41 | Antibacterial and cytotoxic effects | [34] |
| Plants | | | | |
| Apple extract | Ag | 22–30 | Great antibacterial effects against *Geobacillus stearothermophilus*, *Staphylococcus aureus*, *Pseudomonas aeruginosa*, and *Klebsiella pneumoniae* | [35] |
| *Lavandula vera* | Zn | 60–80 | Valuable antibacterial and anti-biofilm activity | [36, 37] |
| *Psidium guajava* | Se | Spherical/8–20 | Antibacterial effects | [38] |
| *Cassia alata* | ZnO | 60–80 | Antibacterial effect against *Escherichia coli* | [39] |
| *Gnidia glauca and Plumbago zeylanica* | Cu | Spherical/1–5 | Good antibacterial effects | [40] |
| *Andrographis paniculata* | Ag | 54 | Good antifungal activity | [41] |
| *Cassia fistula* | Au | 55–98 | Hypoglycemia treatment | [42] |
| Enzyme and other biological agents | | | | |
| Melanin | Cu | Spherical/66 | Good antibacterial activity against *E. coli* and *L. monocytogenes* | [43] |
| Horseradish peroxidase | Au | 10 | Detection of low concentrations of phenylhydrazine | [44] |
| Macerating enzymes | Ag | Hexagonal/38 | High antibacterial effects | [45] |

**Table 2.**
*Some examples of biologically produced NPs and their corresponding special characteristics.*

## 3. NP biomedical applications

With respect to special properties of NPs discussed before, they have various applications. Here we investigate some of these applications with special look at their uses in biomedical fields.

### 3.1 Drug delivery

NPs are of great interest for being used as a device for site-specific drug delivery with optimum dosage drug release. Current NP-based drug delivery approaches focused mainly on enhancing drug shelf life though improving drug uptake efficiency [46]. NP-based drug carriers are able to cross the blood-brain barrier and tight junctions of the skin epithelial tissue [47]. Also they improve hydrophobic molecule solubility and increase stability of biological therapeutic agents.

NPs enabled us to deliver drugs by various routes including nasal mucosa and oral administration, aerosol method, and topical vaccination. The aerosol technology is used for respiratory disorder drug delivery. Target drug delivery approaches using magnetic NPs are widely being used for cancer therapy, gene therapy, MRI, and cell sorting [48, 49]. For instance, $Fe_3O_4$, $\gamma$-$Fe_2O_3$, and super magnetic iron oxide NPs (SPIONs) are the main NPs used for site-specific drug delivery. The surface properties and particular shape of fullerenes and carbon nanotubes make them attractive for drug delivery. These particles are such small that can pass through cell membrane and deliver agents like DNA and protein into the cells [50, 51].

### 3.2 Antibacterial agent

The prevalence of antibiotic-resistant bacteria species becomes a threat for human health. NPs with significant antibacterial properties and no bacterial resistance are the best alternative for common antibiotics [52]. Ag NPs are the leading NP-based antibacterial agents with significant bactericidal effects on both Gram-negative and Gram-positive bacteria [53]. Every day various NPs with different physicochemical properties and bactericidal activities have been developed, and their mechanism of action and potential side effects are under investigations. Also application of common antibiotics such as ampicillin, chloramphenicol, and kanamycin in the presence of NPs demonstrated the positive effects of this combination. Previous studies showed that NPs can be used as a vehicle for antibiotic delivery. The attachment of NPs to the bacterial surface and induction of damages are reported as the main mechanism bacterial death with NPs [54, 55]. Interaction of NPs with bacterial cell membrane and disruption of its normal function are the most common way of NP bacterial killing. NPs are also able to hindrance bacterial biofilm formation. Furthermore, NPs are able to produce different types of ROS species. For example, Mg NPs are able to produce $O^{2-}$, and ZnO NPs produce $H_2O_2$ and OH. These ROS species interact with bacterial cells and cause acute stress reactions and finally lead to acute microbial death [56, 57].

### 3.3 Biosensor

The optical and electronic properties of NPs make them suitable for biosensor application. The size, type of ion, and shape of NPs are critical parameters affecting SPR peaks and line widths of sensor. The noble metals like Au, Ag, and Pt NPs showed special physicochemical features which make them the most popular components of NP-based biosensors [58]. NPs have different roles in any types of biosensors. For instance, electrochemical biosensor is performed by fixing the potential at a suitable value and determining the current changes versus time. The role of NPs in this type of biosensor is to improve sensitivity and signal detection [59]. In optical biosensors, the free electron oscillation in conduction bands of some metals (Ag, Au, Cu) interacts with light photons and produces a polariton. Size tuning of plasmonic metals is a way for enhancing surface plasmon resonance and making the device suitable for biomedical applications. Using NPs leads to reaching to highest detection sensitivity. Au NPs because of their easy functionalization and showing different colors based on their size and shape are good choices for colorimetric biosensor, plasmonic sensing, immune sensors, and electrochemistry [60]. The Au NPs showed unique stability compared to other metals when used for bio-conjugation production and have valuable sensitive plasmon change which lead to their wide use in classic immunoassays. The stronger Raman and fluorescence enhancement of Ag NPs than Au NPs resulted to their broad uses in optical applications [61]. Also they can easily be oxidized and be used in electrochemical sensors. Ag NPs with ability to detect proteins have been used for cancer detection. Furthermore they were also used for detecting glucose, DNA, dopamine, ascorbic acid, and several other biological molecules. Magnetic NPs are used in sensors through three main approaches including pre-concentration of analyte, magnetic tags, and integration into transducer materials [62, 63].

### 3.4 Diagnostic agent

The special features of NPs such as fluorescence properties, optical scattering and electromagnetic field enhancement, and even transferring light energy to heat resulted in a wide application of these compounds in medical diagnostic field. Furthermore, NPs are excellent carriers for delivery of active biomedical agents. Biomedical imaging is one of the useful tools for human disease diagnosis. The NPs with special optical, magnetic, and radioactive properties can enhance the quality of imaging. It is possible to functionalize NPs with multiple modals and by this way minimize the interface between each modality and provide multimodal agent for better imaging [64]. The optical nano-probes can be designed for being used in linear optical imaging with high-emission quantum energy yield and expanded optical capacity. In the case of fluorescent imaging, the degradation of organic dyes (photo bleaching) and metal complexes under light exposure is alleviated with fluorophore-doped silica NPS. These NPs have been used for untargeted imaging of human epithelial cells of the cervix and targeted imaging of cells A549, HeLa, and HepG2 [65, 66]. Phosphorescence imaging using NPs produced images with lower background autofluorescence and scattered excitation light in the spectral range. Magnetic resonance imaging (MRI) used contrast agents for detection of small tumor and lesions in a normal tissue. The NPs with magnetic functionality are used in MRI, and the ones with larger magnetic moment are preferred [67].

### 3.5 Catalytic agent

NPs have been developed for various catalytic applications. The NP catalytic reactions have several advantages including low reaction temperature, light transparency, and easily immobilization on solid supports, for instance, the catalytic activity of Au NPs in degrading methylene blue demonstrated by Khan et al. [68]. Also the effect of geometrical parameters of supported Au NPs on its carbon monoxide oxidation has been evaluated. The NPs with an average diameter of 2 nm and height of six atomic monolayers showed optimum catalytic activity [69]. The Au NPs on amorphous silica support produced by Mukherjee et al. were able to catalyze hydrogenation of cyclohexene [70]. The Ag NPs produced by lychee (*L. chinensis*) extract showed significant photocatalytic activity even after three times of reusing [71].

### 3.6 Wound healing activity

NPs are a suitable wound dressing agent because of their valuable antibacterial, anti-inflammatory effects and ability to accelerate skin reepithelialization. Reports of healing effects of Ag NPs indicated that these nanoscale materials decrease local matrix metalloproteinase and neutrophil apoptosis. Also they showed inhibitory effects on pro-inflammatory cytokines interferon gamma and tumor necrosis factor alpha [72]. The combination of Ag NPs and collagen results in the formation of component with suitable antibacterial activity. Au NPs do not have any antibacterial effects alone, but their combination with biological agents like collagen and gelatin improve their biocompatibility and biodegradability and make them suitable for wound dressing. Au NP antibacterial properties resulted from their interaction with cell membrane and inhibiting ATP synthase which consequently lead to ROS-independent cell death. The reports indicated that the combination of Au NPs, gallate, and epigallocatechin has positive effects in healing of mouse skin wounds through regulation of angiogenesis and anti-inflammatory effects [73]. Pd, Pt, Se, and ZnO are other promising NPs for regenerative medicine and wound healing. The PAPLAL® solution (Toyokose Pharmaceuticals, Japan) (Shibuya et al. 2014) which is a mixture of Pd NPs and Pt NPs showed protective effects against aging-related skin pathologies and normalized the gene expression levels of Mmp2, Has2, TNF-α, and IL-6 in the skin [74].

Zn NPs have valuable antibacterial effects, and its topical application leads to reduction of inflammation and improvement of skin reepithelialization. $TiO_2$ NP wound treatment enhances body fluid coagulation by making interaction with blood proteins. The formation of adherent crust of a nanocomposite improved healing of wound and inhibited infection and inflammation [75].

Also nanotechnology can be used for delivery of active agents with antimicrobial, anti-inflammatory, and healing effects. Curcumin treatment of diabetic wounds leads to significant enhancement in reepithelialization and an increase in fibroblast proliferation of injured tissue. Curcumin NPs not only have higher lifetime than curcumin but also showed valuable antibacterial effects against methicillin-resistant *Staphylococcus aureus* [76]. With respect to molecular chemistry and self-assembly approaches, it is possible to develop peptide NPs with a variety of medical applications.

Polymeric NPs using both biological and synthetic polymers are of great interest for the development of wound dressing compound. Polymeric NPs are able to stimulate cell proliferation through enhancing angiogenesis and reepithelialization. They are able to stimulate the infiltration of inflammatory cells in the initial phase of healing. Furthermore, they are suitable carriers for therapeutic agents including cytokines, growth factors, and antibiotics which make them suitable for being used in treatment of both normal and delayed infectious wounds [77, 78].

## 4. NP toxicological consideration

Regarding the extensive uses of NPs in foods, paper, drug delivery, biosensor, cancer therapy, and imaging, looking for possible toxicity and long-term exposure side effects and finding the mechanism underlying the adverse effects of NPs seem necessary.

Any toxicity induction of NPs is strongly related to NP base material, shape, size, and functional groups coated at their surface. The smaller NPs have a larger specific surface area which in turn leads to higher interaction cell components including nucleic acid, proteins, and carbohydrates. Also the smaller NPs can penetrate better into cells and interact higher with cells. Surface charge of NPs has strong correlation with their interaction with cells and absorbance. It is showed that the NPs with higher positive charge have higher cytotoxicity effects. Also NPs with positive charge are more toxic than negatively charged NPs [79]. The shape of NPs is the other critical parameter which largely affects their cytotoxicity and antiproliferative effects. For instance, the amorphous $TiO_2$ NPs produced higher levels of oxidative stress and cell surface defects than anatase $TiO_2$ NPs. Also the spherical $Fe_2O_3$ NPs had lower cytotoxic effects than rod-shaped ones. Also cytotoxicity is strongly dependent on the type of cells. For instance, although citrate-capped gold NPs were nontoxic to human liver carcinoma and hamster kidney cells, they were severely toxic to human carcinoma lung cells [80].

NPs can easily penetrate into the cells and interact with cells' normal functions. ROS formation and consequently oxidative stress induction are the common side effects of metal NPs. The produced ROS disrupt normal cell function through attacking essential biological molecules including DNA, enzymes, and lipids. Peroxidation of membrane lipids; enhancing calcium entrance; release of calcium from intracellular stores, protein kinase C, and mitogen-activated protein kinase activation; and DNA damages are some of the main changes that lead to cell death after interaction with NPs [81, 82]. Furthermore the risk of early apoptosis upon exposure to some NPs such as ZnO and $TiO_2$ has been demonstrated. Also CuO, NiO, $TiO_2$, $Fe_3O_4$, ZnO, and $Al_2O_3$ NPs can arrest cell cycle and induce apoptosis. It is demonstrated that the phase of cell arrest depends on the type of cell and NPs. G2/M phase arrest is the most common type of cell arrest induced by metallic NPs. The induction of P53 pathway in NCM460 cells and cyclin-dependent kinase 1 downregulation in HaCaT cells after exposure to ZnO and cyclin B1 downregulation in A549 cells by $TiO_2$ have been reported as the main causes of cell proliferation disruption [83, 84]. Many researches have been done on different cell lines and animal models for finding the mechanism of NP toxicity and physiological changes. The high absorption of gold NPs and their aggregation inside cells are probably the main cause of gold NP toxicity.

Argyria is a condition of the skin and other organs' blue-gray discoloration as a result of long-time exposure to high levels of Ag NPs. Irritation, stomach pain, allergic reactions, and inflammation are reported as the main side effects of body exposure to high levels of Ag NPs [85]. $TiO_2$ and ZnO NPs are widely being used in cream and lotions as sunscreen or materials for water- or stain-repellent properties. The cytotoxicity induction of TiO2 NPs through increasing reactive oxygen species and lactate dehydrogenase has been demonstrated [86]. It revealed that UV and visible light irradiation enhanced ZnO NP cytotoxicity power [87]. Also the Zn NPs were produced by microwave-assisted method, and its in vivo cytotoxicity and levels of distribution in different tissues have been evaluated. According to the obtained results, the produced Zn NPs were classified as nontoxic agents with highest distribution in the testis, liver, and brain [37].

Several assays have been developed for the determination of NP toxicity both in vitro and in vivo. Proliferation assay which measures the active cell metabolism is the most popular method for determining the antiproliferative potency of NPs. 3-(4,5-Dimethylthiazol-2-yl)-2,5-diphenyltetrazolium bromide (MTT), XTT, thymidine incorporation, alamar blue, and clonogenic assay are the most popular methods for determining cell proliferation rate [88, 89]. DNA damage and apoptosis induction of NPs which are mainly due to generation of free radicals can be determined by methods like annexin V, comet assay, DNA laddering, and TdT-mediated dUTP-biotin nick end labeling (TUNEL) [90, 91]. NPs are able to interact with cell membrane and lead to cell integrity destruction and cell death. This phenomenon which is known as cell necrosis is mainly measured by neutral red and trypan blue exclusion assays [92, 93]. The in vivo assays including biochemical tests, histopathological analysis, hematology, and NP bio-distribution are also used for finding the effect of NPs on normal function of cells and tissues [94].

## 5. Conclusion

NPs are becoming the spreadable part of medicine, and their uses are increasing every day. They exhibited promising biomedical uses regarding their special redox potentials, small sizes, high surface area, optical scattering, and fluorescence. Due to special biological effects of these compounds including significantly high antibacterial and antiproliferative effects against a wide range cells, their production and surface modification are increasing for reaching more effective agents. Besides they are able to be used as delivery devices for dispensing drugs and biological agents to specific sites. Owing to the advances in generation of multifunctional NPs, application of NP-based platforms is significantly increasing. While all NPs showed some degree of success in laboratory tests and some of them are now on the market, considering their potent environmental and biological side effects is necessary. Although several researches demonstrated the toxicity of different NPs, the cause of toxicity is mainly unknown. Any NP has its special toxicological characteristics, and there is not a comprehensive method for calculation or grading different NP toxicity. Production of NPs through methods with lowest dangerous side products, optimizing the NP production protocols, and doing both in vitro and in vivo tests of toxicity are the main steps toward production of NPs with lowest negative effects on the environment and human health. Short- and long-term toxicities of NPs and their pharmacokinetic and pharmacodynamic tests should be evaluated for FDA approval.

## Author details

Hamid-Reza Rahimi[1]* and Mohsen Doostmohammadi[2]

1 Department of Pharmacology and Toxicology, Faculty of Pharmacy, Kerman University of Medical Sciences, Kerman, Iran

2 Pharmaceutics Research Center, Institute of Neuropharmacology, Kerman University of Medical Sciences, Kerman, Iran

*Address all correspondence to: hamidrrt@yahoo.com; h_rahimi@kmu.ac.ir

## References

[1] Sahoo SK, Labhasetwar V. Nanotech approaches to drug delivery and imaging. Drug Discovery Today. 2003;**5**(8):1112-2110

[2] Taniguchi N. On the basic concept of nanotechnology. In: Proceedings of the International Conference on Production Engineering. Part II (Japan Society of Precision Engineering). Tokyo: Nano Ethics. 1974;**1**:1-2

[3] Agarwal H, Venkat Kumar S, Rajeshkumar S. A review on green synthesis of zinc oxide nanoparticles-an eco-friendly approach. Resource-Efficient Technologies. 2017;**3**(4):406-413

[4] Shin WK, Cho J, Kannan AG, Lee YS, Kim DW. Cross-linked composite gel polymer electrolyte using mesoporous methacrylate-functionalized SiO2 nanoparticles for lithium-ion polymer batteries. Scientific Reports. 2016;**6**(1):26332

[5] Patil G, Khan MI, Patel DK, Sultana S, Prasad R, Ahmad I. Evaluation of cytotoxic, oxidative stress, proinflammatory and genotoxic responses of micro- and nano-particles of dolomite on human lung epithelial cells a 549. Environmental and Toxicological Pharmacology. 2012;**34**(2):436-445

[6] Fröhlich E. The role of surface charge in cellular uptake and cytotoxicity of medical nanoparticle. International Journal of Nanomedicine. 2012;7:5577-5591

[7] Moroz P, Jones SK, Gray BN. Magnetically mediated hyperthermia: Current status and future directions. International Journal of Hyperthermia. 2002;**18**:267-284

[8] Moroz P, Jones SK, Gray BN. Tumor response to arterial embolization hyperthermia and direct injection hyperthermia in a rabbit liver tumor model. Journal of Surgical Oncology. 2002;**80**(3):149-156

[9] Treuel L, Eslahian KA, Docter D, Lang T, Zellner R, Nienhaus K, et al. Physicochemical characterization of nanoparticles and their behavior in the biological environment. Physical Chemistry Chemical Physics. 2014;**16**(29):15053-15067

[10] Li X, Xu H, Chen ZS, Chen G. Biosynthesis of nanoparticles by microorganisms and their applications. Journal of Nanomaterials. 2014;**2011**:1-16

[11] Reverberi AP, Kuznetsov NT, Meshalkin VP, Salerno M, Fabiano B. Systematical analysis of chemical methods in metal nanoparticles synthesis. Theoretical Foundations of Chemical Engineering. 2016;**50**(1):59-66

[12] Konrad A, Herr U, Tidecks R, Kummer F, Samwer K. Luminescence of bulk and nanocrystalline cubic yttria. Journal of Applied Physics. 2001;**90**(7):3516-3523

[13] Tschopp MA, Murdoch HA, Kecskes LJ, Darling KA. Bulk nanocrystalline metals: Review of the current state of the art and future opportunities for copper and copper alloys. Journal of Metals. 2014;**66**(6):1000-1019

[14] Kim M, Osone S, Kim T, Higashi H, Seto T. Synthesis of nanoparticles by laser ablation: A review. Kona Powder and Particle Journal. 2017;**34**:80-90

[15] Carter CB, Norton MG, Carter CB, Norton MG. Sols, gels, and organic chemistry. Ceramic Materials. New York, NY: Springer. 2007:400-411

[16] Abdul Salam H, Sivaraj R, Venckatesh R. Green synthesis and

characterization of zinc oxide nanoparticles from Ocimum basilicum L. var. purpurascens Benth.-Lamiaceae leaf extract. Materials Letters. 2018;**9**(1):015008

[17] Srikar SK, Giri DD, Pal DB, Mishra PK, Upadhyay SN. Green synthesis of silver nanoparticles: A review. Green and Sustainable Chemistry. 2016;**6**(1):34-56

[18] Zhang X, Yan S, Tyagi RD, Surampalli RY. Synthesis of nanoparticles by microorganisms and their application in enhancing microbiological reaction rates. Chemosphere. 2011;**82**(4):489-494

[19] Iravani S. Green synthesis of metal nanoparticles using plants. Green Chemistry. 2011;**13**(10):2638

[20] Harris AT, Bali R. On the formation and extent of uptake of silver nanoparticles by live plants. Journal of Nanoparticle Research. 2008;**10**(4):691-695

[21] Gardea-Torresdey JL, Parsons JG, Gomez E, Peralta-Videa J, Troiani HE, Santiago P, et al. Formation and growth of Au nanoparticles inside live alfalfa plants. Nano Letters. 2002;**2**(4):397-401

[22] Manceau A, Nagy KL, Marcus MA, Lanson M, Geoffroy N, Jacquet T, et al. Formation of metallic copper nanoparticles at the soil-root interface. Environmental Scientific and Technology. 2008;**42**(5):1766-1772

[23] Khan M, Khan M, Adil SF, Tahir MN, Tremel W, Alkhathlan HZ, et al. Green synthesis of silver nanoparticles mediated by Pulicaria glutinosa extract. International Journal of Nanomedicine. 2013;**8**:1507-1516

[24] Ghosh S, Patil S, Ahire M, Kitture R, Gurav DD, Jabgunde AM, et al. Gnidia glauca flower extract mediated synthesis of gold nanoparticles and evaluation of its chemocatalytic potential. Journal of Nanobiotechnology. 2012;**10**(**1**):17

[25] Rautela A, Rani J, Debnath M. Green synthesis of silver nanoparticles from *Tectona grandis* seeds extract: Characterization and mechanism of antimicrobial action on different microorganisms. Journal of Analytical Science and Technology. 2019;**10**:5

[26] Aljabali A, Akkam Y, Al Zoubi M, Al-Batayneh K, Al-Trad B, Abo Alrob O, et al. Synthesis of gold nanoparticles using leaf extract of *Ziziphus zizyphus* and their antimicrobial activity. Nanomaterials. 2018;**8**(3):174

[27] Haverkamp RG, Marshall AT. The mechanism of metal nanoparticle formation in plants: Limits on accumulation. Journal of Nanoparticle Research. 2009;**11**(6):1453-1463

[28] Shakibaie M, Hajighasemi E, Adeli-Sardou M, Doostmohammadi M, Forootanfar H. Antimicrobial and anti-biofilm activities of Bi subnitrate and BiNPs produced by Delftia sp. SFG against clinical isolates of Staphylococcus aureus, Pseudomonas aeruginosa, and Proteus mirabilis. IET Nanobiotechnology. 2019;**13**(4):377-381

[29] Sweeney RY, Mao C, Gao X, Burt JL, Belcher AM, Georgiou G, et al. Bacterial biosynthesis of cadmium sulfide nanocrystals. Chemical Biology. 2004;**11**(11):1553-1559

[30] Arya A, Gupta K, Chundawat TS, Vaya D. Biogenic synthesis of copper and silver nanoparticles using green alga *Botryococcus braunii* and its antimicrobial activity. Bioinorganic Chemistry and Applications. 2018;**2018**:1-9

[31] Kowshik M, Ashtaputre S, Kharrazi S, Vogel W, Urban J, Kulkarni SK, et al. Extracellular synthesis of silver nanoparticles by

a silver-tolerant yeast strain MKY3. Nanotechnology. 2003;**14**(1):95-100

[32] Korbekandi H, Ashari Z, Iravani S, Abbasi S. Optimization of biological synthesis of silver nanoparticles using fusarium oxysporum. Iranian Journal of Pharmaceutical Research. 2013;**12**(3):289-298

[33] Iqtedar M, Aslam M, Akhyar M, Shehzaad A, Abdullah R, Kaleem A. Extracellular biosynthesis, characterization, optimization of silver nanoparticles (AgNPs) using *Bacillus mojavensis* BTCB15 and its antimicrobial activity against multidrug resistant pathogens. Preparative Biochemistry & Biotechnology. 2019;**49**(2):136-142

[34] Shahzad A, Saeed H, Iqtedar M, Hussain SZ, Kaleem A, Abdullah R, et al. Size-controlled production of silver nanoparticles by aspergillus fumigatus BTCB10: Likely antibacterial and cytotoxic effects. Journal of Nanomaterials. 2019;**2019**:1-14

[35] Nthunya LN, Derese S, Gutierrez L, Verliefde AR, Mamba BB, Barnard TG, et al. Green synthesis of silver nanoparticles using one-pot and microwave-assisted methods and their subsequent embedment on PVDF nanofibre membranes for growth inhibition of mesophilic and thermophilic bacteria. New Journal of Chemistry. 2019;**43**(10):4168-4180

[36] Shakibaie M, Alipour-Esmaeili-Anari F, Adeli-Sardou M, Ameri A, Doostmohammadi M, Forootanfar H, et al. Antibacterial and anti-biofilm effects of microwave-assisted biologically synthesized zinc nanoparticles. Nanomedicine Journal. 2019;**6**:13085

[37] Salimi A, Rahimi HR, Forootanfar H, Jafari E, Ameri A, Shakibaie M. Toxicity of microwave-assisted biosynthesized zinc nanoparticles in mice: A preliminary study. Artificial Cells, Nanomedicine, and Biotechnology. 2019;**47**(1):1846-1858

[38] Alam H, Khatoon N, Raza M, Ghosh PC, Sardar M. Synthesis and characterization of nano selenium using plant biomolecules and their potential applications bionanoscience. Bionanoscince. 2019;**9**(1):96-104

[39] Happy A, Soumya M, Venkat Kumar S, Rajeshkumar S, Sheba Rani ND, Lakshmi T, et al. Phyto-assisted synthesis of zinc oxide nanoparticles using *Cassia alata* and its antibacterial activity against *Escherichia coli*. Biochemistry and Biophysics Reports. 2019;**17**:208-211

[40] Jamdade DA, Rajpali D, Joshi KA, Kitture R, Kulkarni AS, Shinde VS, et al. *Gnidia glauca* - and *Plumbago zeylanica* -mediated synthesis of novel copper nanoparticles as promising antidiabetic agents. Advanced in Pharmaceutical Science. 2011;**2011**:1-5

[41] Kotakadi VS, Gaddam SA, Subba Rao Y, Prasad TNVKV, Varada Reddy A, Sai Gopal DVR. Biofabrication of silver nanoparticles using *Andrographis paniculata*. Europian Journal of Medical Chemistery. 2014;**12**(73):135-140

[42] Daisy P, Saipriya K. Biochemical analysis of *Cassia fistula* aqueous extract and phytochemically synthesized gold nanoparticles as hypoglycemic treatment for diabetes mellitus. International Journal of Nanomedicine. 2012;**7**:1189-1202

[43] Roy S, Rhim J-W. Melanin-mediated synthesis of copper oxide nanoparticles and preparation of functional agar/CuO NP nanocomposite films. Journal of Nanomaterials. 2019;**2019**:1-10

[44] Parashar A, Sachin Kedare P, Alex SA, Chandrasekaran N, Mukherjee A. A novel enzyme-mediated gold

nanoparticle synthesis and its application for: In situ detection of horseradish peroxidase inhibitor phenylhydrazine. New Journal of Chemistry. 2017;**41**(24):15079-15086

[45] Thapa R, Bhagat C, Shrestha P, Awal S, Dudhagara P. Enzyme-mediated formulation of stable elliptical silver nanoparticles tested against clinical pathogens and MDR bacteria and development of antimicrobial surgical thread. Annals of Clinical Microbiology and Antimicrobials. 2017;**16**:39

[46] Farokhzad OC, Langer R. Impact of nanotechnology on drug delivery. ACS Nano. 2009;**3**(1):16-20

[47] Zhou Y, Peng Z, Seven ES, Leblanc RM. Crossing the blood-brain barrier with nanoparticles. Journal of Controlled Release. 2018;**28**(270):290-303

[48] Estelrich J, Sánchez-Martín MJ, Busquets MA. Nanoparticles in magnetic resonance imaging: From simple to dual contrast agents. International Journal of Nanomedicine. 2015;**10**:1727-1741

[49] Fan TX, Chow SK, Zhang D. Biomorphic mineralization: From biology to materials. Progress in Materials Science. 2009;**54**(5):542-659

[50] Guan X, Avci-Adali M, Alarçin E, Cheng H, Kashaf SS, Li Y, et al. Development of hydrogels for regenerative engineering. Biotechnology Journal. 2017;**12**(5):1600394

[51] Aboofazeli R. Carbon nanotubes: A promising approach for drug delivery. International Journal of Pharmaceutical Research. 2019;**9**:1-3

[52] Kumar M, Curtis A, Hoskins C. Application of nanoparticle technologies in the combat against anti-microbial resistance. Pharmaceutics. 2018;**14**(10):11

[53] Abbaszadegan A, Ghahramani Y, GholamiA,HemmateenejadB,DorostkarS, Nabavizadeh M, et al. The effect of charge at the surface of silver nanoparticles on antimicrobial activity against gram-positive and gram-negative bacteria: A preliminary study. Journal of Nanomaterials. 2015;**2015**:1-8

[54] Nirmala Grace A, Pandian K. Antibacterial efficacy of aminoglycosidic antibiotics protected gold nanoparticles-a brief study. Colloids and Surfaces A: Physicochemical and Engineering Aspects. 2007;**297**:63-67

[55] Li WR, Xie XB, Shi QS, Zeng HY, Ou-Yang YS, Ben CY. Antibacterial activity and mechanism of silver nanoparticles on *Escherichia coli*. Applied Microbiology and Biotechnology. 2010;**85**(4):1115-1122

[56] Malka E, Perelshtein I, Lipovsky A, Shalom Y, Naparstek L, Perkas N, et al. Eradication of multi-drug resistant bacteria by a novel Zn-doped CuO nanocomposite. Small. 2013;**9**(23):4069-4076

[57] Li Y, Zhang W, Niu J, Chen Y. Mechanism of photogenerated reactive oxygen species and correlation with the antibacterial properties of engineered metal-oxide nanoparticles. ACS Nano. 2012;**6**(6):5164-5173

[58] Wu B, Kuang Y, Zhang X, Chen J. Noble metal nanoparticles/carbon nanotubes nanohybrids: Synthesis and applications. Nano Today. 2011;**6**(1):75-90

[59] Ding L, Bond AM, Zhai J, Zhang J. Utilization of nanoparticle labels for signal amplification in ultrasensitive electrochemical affinity biosensors: A review. Analitica Chimica Acta. 2013;**797**:1-12

[60] Aldewachi H, Chalati T, Woodroofe MN, Bricklebank N,

SharrackB, GardinerP. Goldnanoparticle-based colorimetric biosensors. Nanoscale. 2018;**10**(1):18-33

[61] Farka Z, Juřík T, Kovář D, Trnková L, Skládal P. Nanoparticle-based immunochemical biosensors and assays: Recent advances and challenges. Chemical Reviews. 2017;**117**(15):9973-10042

[62] Shin JW, Yoon J, Shin M, Choi JW. Electrochemical dopamine biosensor composed of silver encapsulated MoS2 hybrid nanoparticle. Biotechnology and Bioprocess Engineering. 2019;**24**(1):135-144

[63] Salahandish R, Ghaffarinejad A, Naghib SM, Niyazi A, Majidzadeh A-K, Janmaleki M, et al. Sandwich-structured nanoparticles-grafted functionalized graphene based 3D nanocomposites for high-performance biosensors to detect ascorbic acid biomolecule. Scientific Reports. 2019;**9**(1):1226

[64] Chen G, Roy I, Yang C, Prasad PN. Nanochemistry and nanomedicine for nanoparticle-based diagnostics and therapy. Chemical Reviews. 2016;**116**(5):2826-2885

[65] Yan H, Wang J, Yi P, Lei H, Zhan C, Xie C, et al. Imaging brain tumor by dendrimer-based optical/paramagnetic nanoprobe across the blood-brain barrier. Chemical Communities. 2011;**47**(28):8130

[66] Kumar R, Roy I, Ohulchanskyy TY, Goswami LN, Bonoiu AC, Bergey EJ, et al. Covalently dye-linked, surface-controlled, and bioconjugated organically modified silica nanoparticles as targeted probes for optical imaging. ACS Nano. 2008;**2**(3):449-456

[67] Kumar R, Ohulchanskyy TY, Roy I, Gupta SK, Borek C, Thompson ME, et al. Near-infrared phosphorescent polymeric nanomicelles: Efficient optical probes for tumor imaging and detection. ACS Applied Materials & Interfaces. 2009;**1**(7):1474-1481

[68] Khan S, Alam F, Azam A, Khan AU. Gold nanoparticles enhance methylene blue-induced photodynamic therapy: A novel therapeutic approach to inhibit *Candida albicans* biofilm. International Journal of Nanomedicine. 2012;7:3245-3257

[69] Saint-Lager MC, Laoufi I, Bailly A, Robach O, Garaudée S, Dolle P. Catalytic properties of supported gold nanoparticles: New insights into the size-activity relationship gained from in operando measurements. Faraday Discussions. 2011;**152**:253-265

[70] Mukherjee P, Patra CR, Ghosh A, Kumar R, Sastry M. Characterization and catalytic activity of gold nanoparticles synthesized by autoreduction of aqueous chloroaurate ions with fumed silica. Chemistry of Materials. 2002;**14**(4):1678-1684

[71] Khan AU, Yuan Q, Wei Y, Khan ZUH, Tahir K, Khan SU, et al. Ultra-efficient photocatalytic deprivation of methylene blue and biological activities of biogenic silver nanoparticles. Journal of Photochemistry and Photobiology, B: Biology. 2016;**159**:49-58

[72] Shin SH, Ye MK, Kim HS, Kang HS. The effects of nano-silver on the proliferation and cytokine expression by peripheral blood mononuclear cells. International Immunopharmacology. 2007;**15**(7):1813-1818

[73] Chen SA, Chen HM, YDer Y, Hung CF, Tu CS, Liang YJ. Topical treatment with anti-oxidants and Au nanoparticles promote healing of diabetic wound through receptor for advance glycation end-products. Europian Journal of Pharmaceutical Sciences. 2012;**18**(47):875-883

[74] Shibuya S, Ozawa Y, Watanabe K, Izuo N, Toda T, Yokote K, et al. Palladium and platinum nanoparticles attenuate aging-like skin atrophy via antioxidant activity in mice. PLoS One. 2014;**9**(10):e109288

[75] Seisenbaeva GA, Fromell K, Vinogradov VV, Terekhov AN, Pakhomov AV, Nilsson B, et al. Dispersion of TiO2 nanoparticles improves burn wound healing and tissue regeneration through specific interaction with blood serum proteins. Scientific Reports. 2017;7(1):15448

[76] Kleinman HK, Philp D, Hoffman MP. Role of the extracellular matrix in morphogenesis. Current Opinion in Biotechnology. 2003;**14**(5):526-532

[77] Heit YI, Dastouri P, Helm DL, Pietramaggiori G, Younan G, Erba P, et al. Foam pore size is a critical interface parameter of suction-based wound healing devices. Plastic and Reconstructive Surger. 2012;**129**(3):589-597

[78] Jiang B, Larson JC, Drapala PW, Pérez-Luna VH, Kang-Mieler JJ, Brey EM. Investigation of lysine acrylate containing poly(N-isopropylacrylamide) hydrogels as wound dressings in normal and infected wounds. Journal of Biomedical Materials Research. Part B, Applied Biomaterials. 2012;**100**(3):668-676

[79] Baek M, Kim MK, Cho HJ, Lee JA, Yu J, Chung HE, et al. Factors influencing the cytotoxicity of zinc oxide nanoparticles: Particle size and surface charge. Journal of Physics Conference Series. 2011;**304**(1):012044

[80] Patra HK, Banerjee S, Chaudhuri U, Lahiri P, Dasgupta AK. Cell selective response to gold nanoparticles. Nanomedicine. 2007;**3**(2):111-119

[81] Huang YW, Cambre M, Lee HJ. The toxicity of nanoparticles depends on multiple molecular and physicochemical mechanisms. International Journal of Molecular Sciences. 2017;**18**(12):2702

[82] Shakibaie M, Abharian A, Forootanfar H, Ameri A, Jafari M, Reza Rahimi H. Cytotoxicity investigations of biogenic tellurium nanorods towards PC12 cell line. IET Nanobiotechnology. 2018;**12**(8):1144-1149

[83] Hanagata N, Zhuang F, Connolly S, Li J, Ogawa N, Xu M. Molecular responses of human lung epithelial cells to the toxicity of copper oxide nanoparticles inferred from whole genome expression analysis. ACS Nanotechnology. 2011;**5**(12):9326-9338

[84] Kansara K, Patel P, Shah D, Shukla RK, Singh S, Kumar A, et al. TiO2 nanoparticles induce DNA double strand breaks and cell cycle arrest in human alveolar cells. Environmental and Molecular Mutagenesis. 2015;**56**(2):204-217

[85] White JML, Powell AM, Brady K, Russell-Jones R. Severe generalized argyria secondary to ingestion of colloidal silver protein. Clinical and Experimental Dermatology. 2003;**28**(3):254-256

[86] Jin CY, Zhu BS, Wang XF, Lu QH. Cytotoxicity of titanium dioxide nanoparticles in mouse fibroblast cells. Chemical Research in Toxicology. 2008;**21**(9):1871-1877

[87] Yang Q, Ma Y. Irradiation-enhanced cytotoxicity of zinc oxide nanoparticles. International Journal of Toxicology. 2014;**33**(3):187-203

[88] Sayes CM, Reed KL, Warheit DB. Assessing toxicology of fine and nanoparticles: Comparing in vitro measurements to in vivo pulmonary toxicity profiles. Toxicological Sciences. 2007;**97**(1):163-180

[89] Punshon G, Vara DS, Sales KM, Kidane AG, Salacinski HJ, Seifalian AM. Interactions between endothelial cells and a poly(carbonate silsesquioxane- bridge-urea)urethane. Biomaterials. 2005;**26**(32):6271-6279

[90] Plackal Adimuriyil George B, Kumar N, Abrahamse H, Ray SS. Apoptotic efficacy of multifaceted biosynthesized silver nanoparticles on human adenocarcinoma cells. Scientific Reports. 2018;**8**(1):14368

[91] Perde-Schrepler M, Florea A, Brie I, Virag P, Fischer-Fodor E, Vâlcan A, et al. Size-dependent cytotoxicity and genotoxicity of silver nanoparticles in Cochlear cells in vitro. Journal of Nanomaterials. 2019;**2019**:1-12

[92] Hu W, Culloty S, Darmody G, Lynch S, Davenport J, Ramirez-Garcia S, et al. Neutral red retention time assay in determination of toxicity of nanoparticles. Marine Environmental Research. 2015;**111**:158-161

[93] Schneider T, Westermann M, Glei M. In vitro uptake and toxicity studies of metal nanoparticles and metal oxide nanoparticles in human HT29 cells. Archive for Toxicology. 2017;**91**(11):3517-3527

[94] Kumar V, Sharma N, Maitra SS. In vitro and in vivo toxicity assessment of nanoparticles. International Nanotechnology Letter. 2007;**7**(4):243-256

Chapter 2

# Role of Nanobiotechnology in Drug Discovery, Development and Molecular Diagnostic

*Deepak Kumar Dash, Rajni Kant Panik, Anil Kumar Sahu and Vaibhav Tripathi*

## Abstract

Nano-biotechnology has already tested its magnitude in a number of sections of existence science and biotechnology field. It is no longer hyperbole to say that in future, nano-scale method would in reality take the associated science area to the subsequent level. Since, there are technical hurdles present; despite the fact that scientists are giving their great to overcome such problems. Applications of nano-biotechnology have already been discussed in this chapter. Future potential are really associated with innovative amendment of such applications. Despite of some impedance, this technology presents giant hope in the future. It performs most important position in distinct sorts of biomedical application such as shipping of drug, gene therapy, biosensors, biomarkers and molecular imaging. It additionally leads to innovations in this field. The fundamental lookup goal of this discipline would be the innovation of early analysis approach and cure with target-specific remedy therapy. Although there would possibly be some safety worries with admire to the in vivo use of nanoparticles, research are in region to decide the nature and extent of adverse events.

**Keywords:** nanobiotechnology, drug discovery, drug design, nanoparticle, biosensors

## 1. Introduction

The effective treatment for disease requires the improvement of diagnostic method, development of optimized drug loaded formulation and incorporation of optimized formulation in to suitable delivery system in terms of role of nanotechnology in drug discovery and development.

There has been challenging task for researchers to increase analytical capacity with improved data quality, consuming less sample volume for storage and screening of cell and tissue library at molecular level. In current scenario the innovative format of nanotechnology offers advancement in the technology, overcoming the initial challenges of unreliable data, consuming high sample and various other issues [1].

Pharmaceutical industries are continually facing a challenge to find better drug discovery technologies because of availability of various competitive other medicinal market. This industry requires discovering and developing new medicines for the effective treatment of wide range of diseases likely to grow regulatory challenges, costing revolution and various other barriers. Recently, all pharmaceutical companies giving importance to common processes including cloning and expressing human receptors and enzymes for discovering innovative drugs, that allow high throughput, automated screening and the application of combinatorial chemistries. Presently, the drug discovery industry should have massive amounts of data about life's molecular components in the area of genomics and proteomics revolution [2].

The Pharmaceutical and biotechnology firms have spent billions of dollars on novel technologies in order to improve productivity, accelerate the drug discovery and development processes and sustain market share.

Nanotechnology refers to development of innovative research and technology at the atomic, molecular and macromolecular level with controlled manipulation, and the study of structures and devices are done in the 1 to 100 nanometers range. The conclusive remark at this scale, Nanoparticle take on beneficial novel properties and functions like small size, surface tailorability, improved solubility, and multifunctionality as compared to seen in the bulk scale. This nanofield may open many new avenues of research for biologists. The novel nanotechnology using nano-materials can interact with complex biological organ consuming low level of biomolecules.

The development and design of multifunctional nanoparticle offers new opportunity for interdisciplinary researchers that can target, diagnose and treat diseases such as cancer. Within the field of discovery and development of nanomaterials, nanotechnology focuses to improve diagnostic methods, improved drug loaded delivery system for the effective and enhanced therapy. Now a days, the scientific community is paying attention on the physical, chemical and biological properties of nano-sized materials so that development of new applications can be done in order to improve human health.

Earlier the nanotechnological concepts have been discussed in 1959 by physicist Richard Feynman during his talk "There's Plenty of Room at the Bottom," describing manipulation of atom is a key to possible different method of synthesis. In 1974, Norio Taniguchi used the term "nanotechnology." The scanning tunneling microscope was invented in the year 1981, ensures the clear visibility of individual atoms and bonds. Initially the term nanotechnology was used in designing nanoscale devices studying about carbon nanotube [3]. The National Nanotechnology Initiative (NNI) defines nanotechnology as research and development of nanosized delivery system with novel functional properties. Subsequently manipulation of atoms could be possible to create efficient nanomaterials with characterized *invitro* and *invivo* parameters. Thus the inherent nanoscale functional properties to the biological tissues, proves that nanotechnology could be feasible for the application to the life sciences successfully. With growing research in the field of nanotechnology, large number of nanotechnologies using nanomaterials has been studied, but there is no study about safety and toxicity. For this purpose *in vitro* diagnostic use of nanotechnology without any safety risks to people and *in vivo* characterization of nanoparticles, ranging smaller size $<50$ nm in size, were studied. The smaller size of nanoparticle concerns over the entry and localization of the particles in to the cells, but there are still many unanswered questions related to huge diversity of materials used and its wide range in the sizes of nanoparticles, these effects will vary a lot. It has been conceived by many researchers that particular sizes might turn out to have toxic effects and consequently further investigation are required. The FDA approval is mandatory for clinically approval the applications of nanotechnology and substantial regulatory problems in the nanotechnology-based product. Consequently the term "nanobiotechnology," has come in to existence as a unique fusion of biotechnology and nanotechnology [4]. This article will provide an integrated overview of application of nanobiotechnology based molecular diagnostics, drug discovery, and drug delivery in the development of nanomedicine with the relationships.

## 2. Nanobiotechnology

The word "Nanobiotechnology" is a combined study of "The Nanotechnology" including design, development and application of nanomaterials & devices and "The Biotechnology" including the various function of biological site like micro-organisms [4]. The field of nanobiotechnology will grow in the development and discovery of drug in upcoming scenario at exponential rate, Where atom or molecule level devices can be constructed by incorporation of drug in to suitable biocompatible delivery system. Hence nanobiotechnology can help various aspects of biological problem with the help of nanotechnology and information technology. This technology has capacity to build bridge among different branches of sciences providing newer challenges and opening new door in the field of research & diagnostics, education in the near future [5]. Relation between nanobiotechnology, bionanotechnology, and nanobiology are the terms referring to the intersection of nanotechnology and biology. Nanobiotechnology includes recent advances in nanotechnology to improve biotechnology, and bionanotechnology, which aims to have advantage of natural/biomimetic approaches to create devices with nanosize range [6].

This article will focus on the principal trends and the implications of nanobiotechnology in drug discovery, development and molecular diagnostic.

## 3. Advantages of nanobiotechnology

Nanoformulation based on oral and topical drug delivery in biotechnology can stimulate molecular imaging for brain tissue engineering and drug delivery through blood brain barrier, Improvement of CNS directed neuroprotection or regeneration, drug uptake assistance, showing fewer side-effects compared to conventional molecules, more active therapeutically in treating brain tumors, Successful biodistribution, High bioavailability, DNA and RNA molecules provide high affinity and binding on specific target site, devoid of enzymatic degradation enriches lymphatic

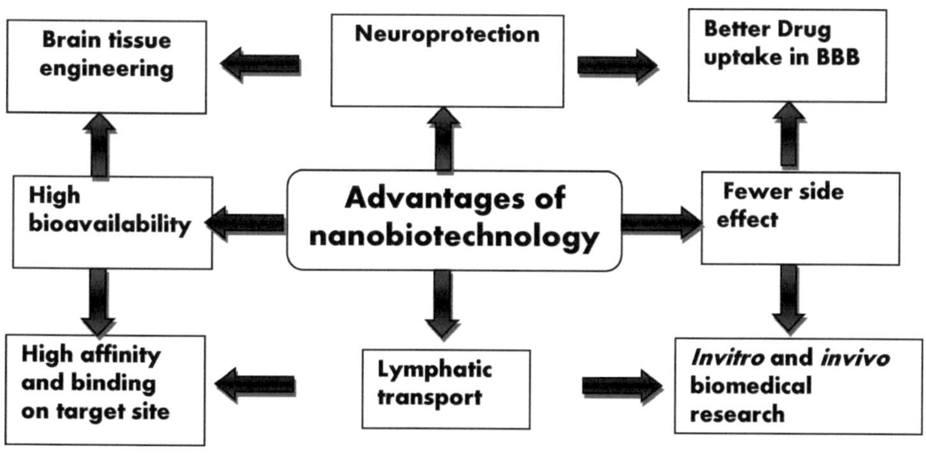


**Figure 1.**
*Advantages of nanobiotechnology.*

transport, development of biosensors, nontoxic and biodegradable nature, enhancement of the potency of anticancer drugs, in vivo and in vitro biomedical research and accomplishment of easy Drug targeting (**Figure 1**) [7–13].

## 4. Nanobiotechnology in drug discovery and development

The application of nanobiotechnology has an impact on diagnostics and drug delivery. Currently, researchers are starting to use nanotechnology in the field of drug discovery and development. Nanocrystals (QDs) and other nanoparticles (gold colloids, magnetic nanoparticles, nanobarcodes, nanobodies, dendrimers, fullerenes, and nanoshells) have unique advantages for the development of drug and its discovery i.e. QDs and magnetic nanoparticles are being used for barcoding of specific analytes which are key components of the bio-barcode assay, can be future alternative to the protein chain reaction [1].

The term "Nanomolecular diagnostics" is the use of nanobiotechnology at molecular level to diagnose and treat chronic diseases", extending the limits of molecular diagnostics to the nanoscale. The nanomolecular diagnostic field can meet the rigorous demands of the clinical laboratory and can be cost-effectiveness [6].

Nanoparticles have emerged as promising nanoplatforms for efficient diagnostics and therapeutics possessing characteristic properties at the nanometric scale. The feasible immobilization of specific ligands on the surface of biological site have become ideal candidates for molecularly sensitive detection, molecular imaging, and novel carriers for targeted drug and gene delivery, targeted photothermal therapy [6]. The nanoparticle in the form of nano-vesicle surrounded by a membrane or a layer holds convenient surface i.e. spherical, cylindrical, plate-like for molecular assembly of inorganic or polymeric materials and drug. The size and size distribution of nanoparticle becoming extremely critical and plays very significant role for penetration through a pore structure of a cellular membrane. The Biomarkers for occupying their distinguished color requires efficient fluorescent probes which depends upon the narrow distribution of average particle sizes. These narrow sizes particles emit a very wide range of wavelengths. The multifunctional nano particles core with several luminescent layers like magnetic nanoparticles can detect and manipulate the particles. Thus the Nanoparticles provides combine platform of biotechnology, nanotechnology and information technology facilitating molecular, biochemical and biological processes, e.g. genetics and pharmacogenomics.

The surface morphology of nanoparticles can be redesigned to reach at their target like blood brain barrier and dermal tight junctions more efficiently, overcoming the efficacy issues of drug on the physiological barriers. In case of malignant tumors Nanoparticles with leaky vasculature structure may penetrate the lesion. Nanoparticles may contain natural and synthetic polymer for coating an inorganic core of superparamagnetic materials. Some examples of nanoparticles as contrast agents in magnetic resonance for molecular diagnosis are given here [5].

### 4.1 Gold nanoparticles

Gold nanoparticles are the most typically used nanomaterial in diagnostics; they have many different uses as properly. They may be extensively utilized as a connecting factor and to construct biosensors to hit upon ailment DNA. Rather than a fluorescent molecule, a gold nanoparticle can be attached to an antibody and other molecules, consisting of DNA, can be brought to the nanoparticle to supply bar codes. The gold particles are very chemically reactive and make excellent catalysts at nanometer scale. Recently, gold nanostructures have found a very important role in a wide variety of applications, including bio-imaging, drug delivery, toxic gas detection, biosensors and to demonstrate multiphoton absorption-precipitated luminescence (mail) [14].

### 4.2 The quantum dots for drug discovery

Quantum dots (QDs) are regarded as a unique class of fluorescent labels, with unique optical properties such as high brightness and long-term colloidal and optical stability; these are suitable for optical imaging, drug delivery and optical tracking, fluorescence immunoassay and other medicinal applications. The unique optical property of QDs allows one to investigate the real-time dynamic events in living cells and such events include interaction between intracellular proteins, the mechanisms of intracellular signal transmission and cell growth. Some benefits and downsides have been investigated more suitable optical residences compared with natural dyes like incredible imaging outcomes using organic dyes, absorption of numerous drug molecules for an extended time frame [15].

### 4.3 Conjugates of nanoparticles and DNA protein

Conjugation Nanoparticle-DNA protein hold great promise in biomedical applications. Many Diverse strategies have been developed to conjugate nanoparticles and DNA proteins to assemble and purify nanoparticle-protein link. First, stable and biocompatible nanoparticles are synthesized. Conjugation of the nanoparticle to the DNA protein is then achieved via two different approaches that do not require heavy chemical modifications or cloning, cysteine-gold covalent bonding, or electrostatic attachment of the nanoparticle to charged groups of the protein. Co-functionalization of the nanoparticle with PEG thiols is recommended to help protein folding. Finally, structural characterization is performed with circular dichroism, as this spectroscopy technique has proven to be effective at examining protein secondary structure in nanoparticle-protein conjugates. In general, functionalization of NPs with proteins such as antibodies can be achieved by direct chemical covalent conjugation or electrostatic interactions. The Semi-artificial conjugates of nucleic acids and proteins by both covalent coupling chemistry, or else through noncovalent biomolecular recognition systems, which include receptor-ligands of complementary nucleic acids for immunological detection assays [15].

### 4.4 Nanochips

The Nano Chip System by utilizing electronically enhanced hybridization of complementary DNA strands, integrates advanced microelectronics and molecular biology into a platform technology with broad commercial applications in the fields of genomic diagnostics and has achieved 100% accuracy in the detection of nanoparticle. This technique helps in investigating DNA sequences or the pairing of separated strands of DNA with complementary DNA strands of the acknowledged collection that act as probes. Currently, DNA chips is known as DNA microarray assays that employs the energy of a digital contemporary that separates DNA probes to unique web sites at the array based on charge and size, then test sample (blood) can be analyzed for identifying DNA sequences via hybridization with these probes [16].

### 4.5 Microfluidics (lab-on-a chip)

Microfluidics is the modern science of fluids on the nanometer scale. The nanodiagnostics involve microfluidic or "lab on a chip" structures, in which the combination of numerous approaches of DNA analysis is mixed on an unmarried chip composed of an unmarried glass and silicon substrate. Academically, it is a subdiscipline of fluid mechanics, as the fundamental equations describing the physics of fluids at larger length scales are identical to the equations underlying microfluidics. In general, the microfluidic biosensor platforms offer numerous advantages compared with other traditional methods such as ultracentrifugation, electrophoresis, nuclear magnetic resonance, chromatographic approaches, etc. for detection of bio-species molecules. This device have capable of measuring aqueous reagent and DNA-containing answers, mixing the solutions together, amplifying or digesting the DNA to shape discrete merchandise, Destiny possibilities for the utility of nanotechnology in healthcare and for the development of personalized remedy seem like amazing [17].

## 5. Nanobiotechnology in designing of drug delivery system

Among the new technologies, nano-biotechnology has evoked considerable interest for application in the pharmaceutical industry. Important applications of nano-biotechnology are in the areas of drug discovery, drug development, and drug delivery, and these are collectively referred to as nano-pharmaceuticals. Nano-biotechnology, particularly the use of nanoparticles, has made significant contributions to drug discovery and development [18]. In addition to the use of nano-biotechnology for drug discovery, some drugs are being developed from nano-materials. Well-known examples of these are dendrimers, fullerenes, and nano-bodies. At present time increasing the use of nano-biotechnology in many pharmaceutical and biotechnology industries is anticipated. In case of drug development from formulate to appropriate dose to administer with optimal delivery systems, nanotechnology is being utilized at all the stages. Nano-biotechnology applications are also involving in diagnosis of diseases. In future it may be possible to computers are connected with nano-biotechnology systems and provide the appropriate and complete knowledge about complete model or an individual cell. This virtual representation might be helpful for researcher or scientist to develop novel drugs with high rate of accuracy and precision without conducting any experiment in living organisms.

The major challenges in front of drug delivery scientists are poor solubility, high molecular size and low bioavailability for clinical candidates. Other challenges in this field are pediatric and geriatric drug administration, protein and peptide drug delivery etc. Today's major demand in drug delivery field is to develop ideal, safe and effective, non-invasive drug delivery methods.

Nano-biotechnology sector play a significant role to overcome the above drug delivery problems. It provides the following solution related drug delivery problems: (a) With the help of this technology particle size of drugs is reduced in nanometer size range, it enhance the surface area and ultimately improve rate of dissolution, (b) Nano-meter size range of drug also useful to improve their solubility, (c) With the help of this technology scientists trying to develop non-invasive routes of drug administration method which can eliminates the use of injectable drugs, (d) developed nanoparticle formulations has better alternate for non-stable and lower shelf lives formulations, (e) nanotechnology based formulations improved the solubility of poorly soluble drug and enhance absorption capability, improved bioavailability and release rate of large molecules, reduced the optimum dose and enhance the safety margin by reducing the side effects, (f) Nano-biotechnology principles help in developing of Sustain and controlled release formulations with better patient compliance [19].

## 6. Nano-biotechnology in molecular diagnostic

Nano technology has gained enormous popularity in recent past decades. In lay man language nano technology is defined as the use of least possible input to get maximum possible output. Nano word is used in terms of size. This technology is considered as wonderful amalgamation of physics and chemistry. When this technique is applied with biology, it brings about a new field termed as Nano-biotechnology; where biochemical processes are modified to get far better results than that of simple bio technological procedures.

As far as the medical field is concerned, the harness of nano- biotechnology for diagnostic purpose is successfully obtained (**Figure 2**). It assists to develop more sensitive diagnostic kits than that of existing one. Such instruments are suitable to probe the bodily problems at cellular pores and receptor level. Moreover, it is due to tiny sized diagnostic materials the degree of toxic and adverse effects have been markedly reduced. New dimensions of diagnostic tools have been explored with the aid of this cutting edge technique. Several modifications are still under developmental phase.

### 6.1 Nano-particles used in the diagnosis

Gold nano-particles are also termed as metallic particles. These are produced from gold salts of either organic, aqueous or both origins. A suitable stabilizer is used to get stable particles with good ligand binding capacity. Usable size of particles is ranging between 3 and 100 nm. The major application is due to their electronic, optical, and thermal properties [20].

Magnetic nano particles are formulated from magnetic materials like $Fe_3O_4$, $Fe_2O_3$, and many other ferrites. Nanoparticles can be incorporated with bio- marker moieties so that they can be utilized to investigate various biomolecules and help in different processes like separation and purification. The involvement of surface coating materials is significant to ascertain the size and kinetics of these particles. That's why the nature of coating materials should be examined before use [21].

Quantum dots are semiconductor nano-crystals that are easy to synthesize and have characteristic properties that are between those of bulk semiconductor and discrete molecules. Their diameter ranges from 2 to 10 nm. Their fluorescent property relies on size of the quantum dot [22].

Carbon nano tubes are composed of graphite. On the basis of number of graphite, these nano tubes are categorized into two classes- tubes with only one layer of graphite

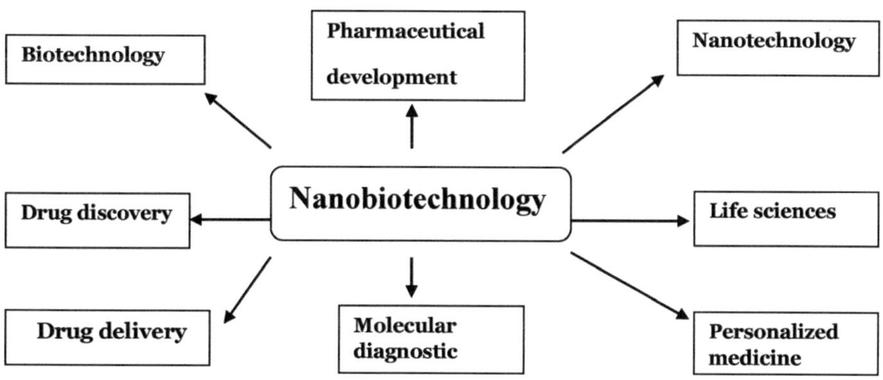


**Figure 2.**
*Application of nanobiotechnology for molecular diagnosis.*

are known as mono walled carbon nano tubes on the other hand, tubes consist of multiple layers of graphite are termed as multi walled carbon nano tubes. The main advantage of such tubes is that they can conduct high electricity with less or no heating effect. This happens because of scattering free traveling of electrons throughout the tube [23].

Liposomes are spherical vesicles where an aqueous core is surrounded by a phospholipid bilayer and cholesterol. The phospholipid consists of a hydrophilic head and two oil-loving tails. The phospholipid that is predominantly used is phosphatidyl choline [24].

Dendrimers are the emerging and well defined polymeric architectures that are known for their large, complex and well defined structures, versatility in drug delivery and high functionality whose properties resemble with biomolecules. These nanostructured macromolecules have potential abilities to entrap and to conjugate the high molecular weight hydrophilic/hydrophobic drugs by host-guest interactions and covalent bonding (prodrug approach) around an inner core respectively. They have different Features like size, shape, branching length, and their surface functionality vary over the design of nanoparticles. For example, Polyamidoamine are usually used dendrimers [25].

Nano bio sensors are applied to investigate biochemical changes with the aid of electronic, optical or magnetic technology. Furthermore, detection and/or quantification of bio molecules like specific base pairs or proteins are also possible. Majority of biosensors work on the principle of key and lock theory or affinity based mechanism; where immobilized tool attaches to target molecule/analyte being sensed in this phenomenon, any change at a localized surface can be examined rather than detecting target in solution. This variation can be sensed by using one of the following methods; viral biosensors, light sensitive biosensors, resonant cantilever and quartz crystal microbalance [26].

## 6.2 Nanotechnology used in diagnosis

### *6.2.1 Nanotechnology-based biochips/microarrays*

A nano material can be perfectly examined by using same sized probe. This hypothesis is applied in molecular diagnosis with the help of biochips or microarrays. This is because the cell organelles/biological moieties exist in nano size and the biochips also falls under in almost similar tiny sized scale. Nano fluidic arrays and protein nano biochips are the examples of nanotechnology based biochips. These chips are capable to isolate and analyze living cell molecules such as genetic material DNA. In future this technique will certainly bring about revived investigation tool for cancer. Nano-fluidic technology is expected to have broad applications in systems biology, personalized medicine, pathogen detection, drug development, and clinical research [27].

### *6.2.2 Nanotechnology-based cytogenetic*

Cytogenetic has been used mainly to elaborate spatial arrangement of the chromosome and screening of abnormalities related to disease. The use of fluorescent in situ hybridization (FISH) is now reaching its limit. Molecular cyto-genetics is now enhanced by use of biomedical nanotechnology, e.g., use of atomic force microscopy (AFM) and quantum dot (QD) FISH [28].

### *6.2.3 Nano-proteomic-based diagnostics*

A comprehensive study including identification, characterization and analysis of a series of bio protein at a specified time is known as proteomics. Proteomics help to recognize the variations which are associated with various pathological conditions.

These variations are identified on the basis of spatio-temporal position of protein molecules. A small change in molecular level can bring about such variations as a consequence proteome may be altered; which are further quantified for clinical diagnosis. Gel- based and gel- free are the common proteomics techniques, which work on separation, observation, analysis and marking the altered proteome [29].

#### *6.2.4 Nanoparticle-based nucleic acid diagnostics*

Screening of pathogenic microorganism on the basis of genomic arrangement has become relevant in current clinical determination. It has been well established central dogma of life that each and every living organism has a peculiar set of DNA, which has been transferred from one generation to next without any sort of alteration; this can also be said to be the individualization by nature. In clinical diagnosis, this DNA fragment may help to check the availability of the microorganism in the test sample. This investigation opens the door for molecular detection. This is carried out by using polymerase chain reaction (PCR) or hybridization techniques, known as amplification method and non-amplification method respectively. Therefore, this technique is considered to be more sensitive and precise than that of other techniques [30].

#### *6.2.5 Nano-bio-sensors*

Bio sensors are sophisticated probes, which are used to investigate the existence and/or concentration of a biological analyte, say for example bio molecule, histology of biological material or any microorganism. Biosensors comprise of following parts: amplifier that detects the analyte and generate a signal along with it a signal transducer that converts the signal into electric impulses and a reader to record and investigate the signal [31].

## 7. Safety issues of nanoparticles

Nano-biotechnology is the fine amalgamation of multi disciplines of science. In pharmaceutical field, the scientists and researchers are trying to overcome the demerits of existing drug molecules with the aid of this wonderful technology. There are many expected potential benefits in bank. In spite of that safety is the main concern with the in- vivo application of nano engineered medicines yet.

It has been revealed that while production or use of nano medicines, such particles can easily get inside the human body and ultimately blood stream and then in various vital organs; where they exert unintended adverse effects, sometimes these effects are fetal for the user.

An in- vivo study on monkeys and rats have disclosed the aggregation of carbon and manganese nano materials in the olfactory bulb, which has produced severe pathological consequences pertaining to nervous and respiratory system.

Several animal studies have shown such sort of abnormal pathological disturbances.

In order to combat with this problem firstly, an efficient probe will have to develop for thorough assessment of risk associated with the use of nano medicine. Second and the most significant challenge would be to develop authentic validation protocol for testing nano medicines in animals so as to get the data of toxic effects in early stage of testing. A restructured, compact and integrated regulatory approach is much needed to look into the expected risks for delivering the promising and safe medicines [2].

## 8. Regulatory perspective of nanobiotechnology

In the event that we talk about the most recent decade, nano bio-innovation has generated some new impedance for the research community, industry, and regulators. Applicable techniques have some loopholes regarding fabrication and processing of nano medicines, that's why special provisions would be required to tide-over such complications [32]. FDA and other regulatory authorities experience many unavoidable consequences for the release of such products. This is due to non-authentic validation program in clinical trial phase [11]. A wing of researchers proposes sturdy contentions, revealing both positive and negative aspects pertaining to the advancement of nano-materials. They suggest that vague protocol throttle research and product development, which represses the development and viability of new nano-medicines. Apart from that, a robust administrative body is made so as to push the business advancement by creating certainty for pharmaceutical firms and trust within the ultimate consumers [33]. Currently, the FDA, EMA (European medical agency), and other regulatory agencies examine each new nanoparticle individually. There is commonly an absence of benchmarks in the assessment of nano-medicines as a unique class of therapeutic agent [34]. Afterwards, administrative corporations might imagine of a complete listing of screening and a powerful approbate system that enclose the complete variety of particle characterization, pharmacology, and toxicology troubles.

## 9. Future prospects

In case we consider in trendy approximately the two major capabilities of nanobiotechnology, first is imaging and diagnostics (quantum dots) and second are proteomics. In proteomics especially study and detect about nucleic acids and proteins [35]. Currently, the Thermal Sensitive drug delivery system has been developed for the treatment of cancer. In this, drug is given in the injection form, which reaches inside the body and converts into insoluble form and accumulates in the tumor cells because the temperature of the tumor cell is more than the normal cell [36]. In spite of the tremendous credibility of nanobiotechology, its use has been considerable due to its toxicity and environmental problems. Despite all this, the drug delivery, drug discovery, gene therapy, molecular imaging, biomarkers, and biosensors fields are exploring possibilities in the nanobiotechnology. Nanobiotechnology has brought a bright future for the clinical diagnosis and targeted drug delivery sector. Now in the field of diagnosis, it has become easier to detect diseased cells rapidly, due to which it is possible to prevent diseased cells from spreading in the body [37]. There is a constant development in this field right now and it is possible that some incurable diseases in the future will actually be cured with the resources of nanobiotechnology.

## 10. Conclusion

Nano-biotechnology is in its primary stage in terms of development. Due to continuous innovative research and abundant use, nano-biotechnology is having a wide impact on the fields of science and technology. Nano-biotechnology is presenting some unique possibilities in medicine, diagnosis and biomedical sciences. The innovation of nano-biotechnology in drug delivery systems has led to the introduction of something new, and with the help of this, it seems possible to treat some incurable diseases. Although the potential for benefits from nano-biotechnology is high, the prospective of nanomedicine is not fully defined. In fact, regulatory bodies do not have appropriate guidelines that balance its risk and safety factors. It would be fair to say that in future nano-biotechnology will play an excellent and unique role in the treatment of human diseases and the study of human physiology. From the research conducted in nano-biotechnology in the last few years, it can be inferred that in future, nano-biotechnology will become an indispensable phase of our daily life.

## Author details

Deepak Kumar Dash*, Rajni Kant Panik, Anil Kumar Sahu and Vaibhav Tripathi
Royal College of Pharmacy, Chhattisgarh Swami Vivekananda Technical University, Bhilai, Chhattisgarh, India

*Address all correspondence to: drdeepak_dash123@yahoo.in

## References

[1] Jain KK. Nanodiagnostics: Application of nanotechnology in molecular diagnostics. Expert Review of Molecular Diagnostics. 2003;**4**:153-161

[2] Elder A, Gelein R, Silva V, Feikert T, Opanashuk L, Carter J. Translocation of inhaled ultrafine manganese oxide particles to the central nervous system. Environmental Health Perspectives. 2006;**114**:1172-1178

[3] Shamshad S, Indira PA. Nanobiotechnology - A new era in science. New Horizons in Biotechnology. 2015;1:208-212

[4] Jain KK. The role of nanobiotechnology in drug discovery. Drug Discovery Today. 2010;**10**(21):1435-1442

[5] Salata OV. Applications of nanoparticles in biology and medicine. Journal of Nanobiotechnology. 2004;**2**:3

[6] Kawadkar J, Chauhan MK, Maharana M. Nanobiotechnology: Application of nanotechnology in diagnosis, drug discovery and drug development. Asian Journal of Pharmaceutical and Clinical Research. 2011;**4**:23-28

[7] Nair SB, Dileep A, Rajanikant GK. Nanotechnology Based Diagnostic and Therapeutic Strategies for Neuroscience with Special Emphasis on Ischemic Stroke. National Institute of Technology Calicut: India; 2011

[8] Sadrieh N, Tyner KM. Nanotechnology and therapeutic delivery: A drug regulation perspective. Therapeutic Delivery. 2010;**1**(1):83-89

[9] Allen TM, Cullis PR. Drug delivery systems: Entering the mainstream. Science. 2004;**303**:1818-1822

[10] Vasir JK, Labhasetwar V. Targeted drug delivery in cancer therapy. Technology in Cancer Research & Treatment. 2005;**4**:363-374

[11] Vasir JK, Reddy MK, Labhasetwar V. Nanosystems in drug targeting: Opportunities and challenges. Current Nanoscience. 2005;**1**:47-64

[12] Feng SS, Mu L, Win KY, Huang G. Nanoparticles of biodegradable polymers for clinical administration of paclitaxel. Current Medicinal Chemistry. 2004;**11**:413-424

[13] Alyautdin RN et al. Significant entry of tubocurarine into the brain of rats by adsorption to polysorbate 80-coated polybutylcyanoacrylate nanoparticles: An in situ brain perfusion study. Journal of Microencapsulation. 1998;**15**:67-74

[14] Ralph AS, Pilar RG, Feng Z, Marco Z, Wolfgang JP. Biological applications of gold nanoparticles. Chemical Society Reviews. 2008;**37**:1896-1908

[15] Niemeyer CM. Semi-synthetic nucleic acid-protein conjugates: Applications in life sciences and nanobiotechnology. Journal of Biotechnology. 2001;**82**(1):47-66

[16] Estes R. Semiconductor packaging technologies advance DNA analysis systems. IVD Technology. 2005;**4**:1-5

[17] O'Connor L. Nanotechnology advances. BioScience. 2005;**4**:1-5

[18] Jain KK. Role of nanobiotechnology in drug discovery. Advances in Experimental Medicine and Biology. 2009c;**655**:37-43

[19] Kumari A, Yadav SK, Yadav SC. Biodegradable polymeric nanoparticles based drug delivery systems. Colloids

and Surfaces B: Biointerfaces. 2010;**75**:1-18

[20] Theron J, Cloete TE, Kwaadsteniet MD. Molecular techniques for determining microbial diversity and community structure in natural environments. Critical Reviews in Microbiology. 2010;**36**:318-339

[21] Gupta AK, Gupta M. Synthesis and surface engineering of iron oxide nanoparticles for biomedical applications. Biomaterials. 2005;**26**:3995-4021

[22] Alivisatos AP. Semiconductor clusters, nanocrystals, and quantum dots. Science. 1996;**271**:933-937

[23] Seetharamappa J, Yellappa S, D'Souza F. Carbon nanotubes next generation of electronic materials. Electrochemical Society Interface. 2006;**15**:23-26

[24] Lasic DD, Lipowsky R, Sackmann E. Structure and dynamics of membranes. Handbook of Biological Physics. 1995;**1**:493-516

[25] Svenson S, Tomalia DA. Dendrimers in biomedical applications - Reflections on the field. Advanced Drug Delivery Reviews. 2005;**57**:2106-2129

[26] Prasad S. Nanobiosensors: The future for diagnosis of disease. Nanobiosensors in Disease Diagnosis. 2014;**3**:1-10

[27] Rajasundari K, Ilamurugu K. Nanotechnology and its applications in medical diagnosis. Journal of Basic and Applied Chemistry. 2011;**1**(2):26-32

[28] Stanley M. Nanodiagnostics: A revolution in biomedical nanotechnology. MOJ Proteomics & Bioinformatics. 2016;**3**(2):34-36

[29] Ahmad Y, Arya A, Gangwar A, Paul S, Bhargava K. Proteomics in diagnosis: Past, present and future. Journal of Proteomics & Genomics. 2014;**1**(1):103

[30] Jamdagni P, Khatri P, Rana JS. Nanoparticles based DNA conjugates for detection of pathogenic microorganisms. International Nano Letters. 2016;**6**:139-146

[31] Alejandro C, Arben M. Nanobiosensors in diagnostics. Nano. 2016;**3**(10):1177

[32] Emerich DF, Thanos CG. Nanotechnology and medicine. Expert Opinion on Biological Therapy. 2003;**3**:655-663

[33] Shaffer C. Nanomedicine transforms drug delivery. Drug Discovery Today. 2005;**10**:1581-1582

[34] Desai N. Challenges in development of nanoparticle-based therapeutics. The AAPS Journal. 2012;**14**(2):282-300

[35] Hamad-Schifferli K et al. Remote electronic control of DNA hybridization throughinductive coupling to an attached metal nanocrystal antenna. Nature. 2002;**415**:152-155

[36] Meyer DE, Shin BC, Kong GA, Dewhirst MW, Chilkoti A. Drug targeting using thermallyresponsive polymers and local hyperthermia. Journal of Controlled Release. 2001;**74**(1-3):213-224

[37] Fakruddin MD, Afroz H, Hossain Z. Prospects and applications of nanobiotechnology: A medical perspective. Journal of Nanbiotechnology. 2012;**10**(31):1-8

Chapter 3

# Nanofibrous Scaffolds for Skin Tissue Engineering and Wound Healing Based on Synthetic Polymers

*Lucie Bacakova, Marketa Zikmundova, Julia Pajorova, Antonin Broz, Elena Filova, Andreu Blanquer, Roman Matejka, Jana Stepanovska, Petr Mikes, Vera Jencova, Eva Kuzelova Kostakova and Alla Sinica*

## Abstract

Nanofibrous scaffolds are popular materials in all areas of tissue engineering, because they mimic the fibrous component of the natural extracellular matrix. In this chapter, we focused on the application of nanofibers in skin tissue engineering and wound healing, because the skin is an organ with several vitally important functions, particularly barrier, thermoregulatory, and sensory functions. Nanofibrous meshes not only serve as carriers for skin cells but also can prevent the penetration of microbes into wounds and can keep appropriate moisture in the damaged skin. The nanofibrous meshes have been prepared from a wide range of synthetic and nature-derived polymers. This review is concentrated on synthetic non-degradable and degradable polymers, which have been explored for skin tissue engineering and wound healing. These synthetic polymers were often combined with natural polymers of the protein or polysaccharide nature, which improved their attractiveness for cell colonization. The nanofibrous scaffolds can also be loaded with various bioactive molecules, such as growth factors, hormones, vitamins, antioxidants, antimicrobial, and antitumor agents. In advanced tissue engineering approaches, the cells on the nanofibrous scaffolds are cultured in dynamic bioreactors enabling appropriate mechanical stimulation of cells and at air-liquid interface. This chapter summarizes recent results achieved in the field of nanofiber-based skin tissue engineering, including results of our research group.

**Keywords:** skin replacements, wound dressings, nanofibers, electrospinning, epidermis, dermis, keratinocytes, fibroblasts, stem cells, vascularization, cell delivery, drug delivery, regenerative medicine

## 1. Introduction

The skin is the largest organ of the human body with several vitally important functions. The skin acts as barrier against adverse effects of the surrounding environment on the organism, such as chemical factors, radiation factors, particularly ultraviolet light, and microbial infection. Other important functions of skin include thermoregulation, sensation of temperature, touch, pressure, and pain, keeping appropriate moisture in the underlying tissues, excretion of ions, water, and various biomolecules (e.g., lipids and proteins), and also production and storage of various biomolecules, such as pigments, vitamin D, and keratins for formation of epidermal appendages (for a review, see [1, 2]). Skin severely and chronically damaged by trauma, burns, bedsores, and by various diseases, e.g., diabetes, cannot exert these functions, which can lead to amputation and even death. Therefore, there is essential need to regenerate the damaged skin, particularly by methods of skin tissue engineering and induction of active wound healing. For these purposes, nanofibrous scaffolds seem to be one of the most promising materials. Nanomaterials in general are defined as features not exceeding 100 nm at least in one dimension, i.e., in diameter in case of nanofibers. However, nanofibers usually used in tissue engineering are often thicker (i.e., several hundreds of nm). In fact, they are submicron-scale fibers, but the term "nanofibers" has become widely used for them. Nanofibers can be obtained by various techniques, such as biological synthesis (e.g., nanocellulose produced by bacteria), self-assembly, phase separation, interfacial polymerization, suitable for electrically conductive materials, melt processing or antisolvent precipitation, and particularly by electrospinning, which has emerged as a relatively simple, elegant, scalable, and efficient technique for fabrication of polymeric nanofibers (for a review, see [2–5]).

The advantage of nanofibrous scaffolds is that they mimic the fibrous component of the natural extracellular matrix (ECM), and therefore they can serve as ECM analogues for tissue engineering. In addition, nanofibrous meshes can act as a protective barrier against penetration of microbes into wounds, can keep the moisture in the damaged skin, and, at the same time, allow gas exchange and absorb the exudate from the wounds. These meshes can also be loaded with various bioactive molecules, such as growth and angiogenic factors, cytokines, hormones, vitamins, antioxidants, antimicrobial and antitumor agents, amino acids (L-arginine), wound healing peptides (e.g., melanocyte-inhibiting factor), and with antimicrobial peptides [6–8]; for a review, see [1, 2]. Therefore, nanofibers can serve not only as tissue engineering scaffolds for skin cells but also as carriers for controlled drug delivery into skin.

The nanofibrous scaffolds for skin tissue engineering and wound healing have been prepared from a wide range of synthetic and nature-derived polymers. Both these groups of polymers contain polymers relatively easily degradable in the human organism, and polymers which are non-degradable or slowly degradable. This review is focused on synthetic polymers, which have been used for creation of nanofibrous scaffolds for skin tissue engineering and wound healing applications. Typical and widely used degradable synthetic polymers include polylactides [9, 10] and their copolymers with polyglycolides [11], or polycaprolactone [6, 12] and its copolymers with polylactides [13]. Examples of biostable synthetic polymers are polyurethane [7], polydimethylsiloxane [14], polyethylene terephthalate [15] or polyethersulfone [16]. Polymers degradable in the human body are suitable as direct scaffolds for skin tissue engineering, while biostable polymers can be rather recommended as "intelligent" wound dressings delivering cells (keratinocytes, fibroblasts or stem cells) and bioactive molecules into wounds.

However, polymers in nanofibrous scaffolds are predominantly used in various combinations—synthetic with natural, degradable with non-degradable—and also in combination with various nanoparticles, e.g., mineral nanoparticles [17], carbon-based nanoparticles [1, 18] or metal-based nanoparticles [19, 20]. The reason of these combinations is to improve the stability, spinnability, wettability, mechanical properties, and bioactivity of nanofibers. The combination of various polymers, nanoparticles, and other components is a strategy commonly used to obtain hybrid materials possessing properties better than those of the individual constituents, regarding their use in scaffolds for tissue engineering or as material for wound dressing [21]. For example, synthetic polymers do not contain adhesion motifs recognizable by cell adhesion receptors, and combination with nature-derived polymers, which are proteins (collagen, gelatin, keratin, fibrin [6, 10, 22, 23]) or polysaccharides (hyaluronic acid, sulfated glycosaminoglycans, such as heparin [24, 25]) can endow them with these motifs, because these polymers are often components of ECM.

In electrospun nanofibrous meshes, the polymers can be combined by various approaches. In blending electrospinning, the polymers are mixed, filled in the same syringe, and electrospun together, which results in creation of fibers with two or more components randomly distributed within a fiber. In multi-jet electrospinning, each polymer is filled in a separate syringe and electrospun individually, which results in creation of meshes with two or more types of nanofibers. These types of nanofibers can be electrospun either concurrently and distributed randomly (i.e., mixing electrospinning) or alternatively and arranged into separate layers (i.e., multilayering electrospinning). In co-axial electrospinning, hybrid nanofibers with a core-shell architecture are created by spinning of two different solutions filled into outer and inner compartments of a co-axial syringe. Finally, an electrospun polymer can be secondarily coated with other polymers or bioactive substances [22, 26–28] (**Figure 1**).

Nanofibrous meshes for skin tissue engineering can also be combined with other material types, such as porous 3D scaffolds or hydrogels in order to reconstruct two main skin layers, i.e., epidermis containing keratinocytes and dermis containing fibroblasts [23, 29–31]. Another advanced approach promising for construction of dermo-epidermal replacements is centrifugal jet spinning, capable of large-scale production of nanofibrous 3D scaffolds [32, 33].

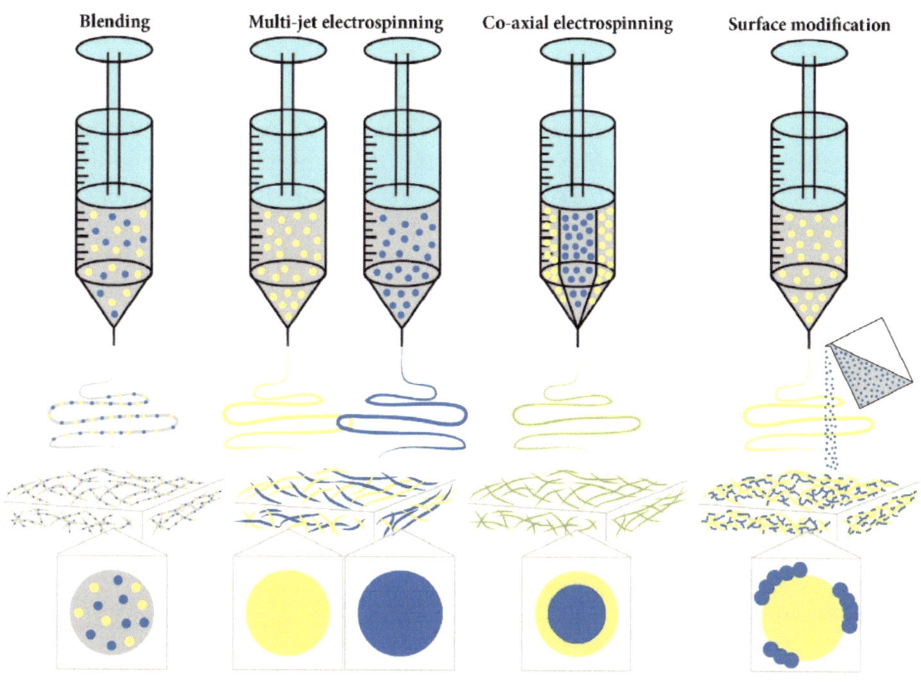


**Figure 1.**
*Modes of combination of various polymers and compounds in nanofibrous scaffolds.*

In the waste majority of studies dealing with skin tissue engineering based on nanofibrous scaffolds, keratinocytes have been cultivated in a conventional static cell culture system, submerged into the culture media, although under physiological condition *in vivo*, keratinocytes are exposed to air. Therefore, in advanced skin tissue engineering, it is necessary to cultivate keratinocytes under appropriate mechanical loading, i.e., strain stress [34] or pressure stress [35], and simultaneously to cultivate them on the scaffolds exposed to the air-liquid interface [34, 36].

This chapter summarizes earlier and recent knowledge on skin tissue engineering and wound dressing applications, based on nanofibrous scaffolds made of synthetic non-degradable and degradable polymers, including our results.

## 2. Nanofibers from synthetic non-degradable polymers

Synthetic non-degradable polymers, explored for creation of electrospun nanofibrous meshes for skin regenerative therapies, included polyurethane, polydimethylsiloxane, polyethylene terephthalate, polyethersulfone, and even polystyrene. This group of polymers also includes non-degradable hydrogels, such as poly(acrylic acid), poly(methyl methacrylate), and poly(di(ethylene glycol) methyl ether methacrylate).

Polyurethane (PU) has been most frequently used from the mentioned polymers, which is due to its elasticity, and also possibility to prepare it in a degradable form, e.g., as poly(ester-urethane) urea (PEUU), which facilitates its applicability in skin tissue engineering [37], while the non-degradable forms of PU (and other non-degradable polymers in general) are rather used in wound dressing applications. Non-degradable PU nanofibrous meshes has been tested as advanced wound dressings loaded with various healing, angiogenic, anti-inflammatory, antioxidant, and antimicrobial substances. For example, blending PU with propolis improved the mechanical strength and hydrophilicity of the nanofibrous membrane, its cytocompatibility with fibroblasts and its antibacterial activity [38]. Blending PU with virgin olive oil endowed the nanofibrous meshes with photoprotective and antioxidant properties [19]. Dextran in composite PU/dextran fibers had angiogenic activity, and also served as a carrier for incorporation of β-estradiol, which accelerated healing of acute cutaneous wounds by its potent anti-inflammatory activity [7]. Another promising nanofibrous membrane applicable for wound dressing was prepared from electrospun PU, treated by plasma and subsequent spraying with chitosan solution containing an inclusion complex of β-cyclodextrin encapsulating berberine, i.e., an isoquinoline alkaloid with antimicrobial and anti-inflammatory activity [39]. Other antimicrobial substances incorporated into PU-based nanofibers included silver nanoparticles [20, 40], copper oxide nanocrystals [19] and antibiotics, such as silver-sulfadiazine [41] and amoxicillin [37]. In general, all these scaffolds showed none or low toxicity towards human HaCaT keratinocytes [20] or fibroblasts [19, 40], and no adverse reactions when implanted into laboratory animals *in vivo* [37, 41]. In addition, copper oxide is known by its angiogenic activity [19]. Nanofibrous meshes created by electrospinning from blends of PU with various concentrations of hydroxypropyl cellulose were also tested for transdermal drug delivery using donepezil hydrochloride, i.e., a drug used for treatment of Alzheimer disease [42]. PU was a component of a novel bilayer wound dressing, consisting of a commercial PU membrane as an outer layer, and an electrospun gelatin/keratin nanofibrous mat as an inner layer. The outer layer acted as a barrier against bacteria and other contaminants, while the inner layers promoted the adhesion, spreading, migration and growth of fibroblasts *in vitro*, and vascularization and wound healing in rats *in vivo* [23].

Polydimethylsiloxane (PDMS) in electrospun nanofibrous scaffolds was used in blends with thermoplastic PU (TPU) in 90:10, 80:20, and 70:30 blend ratios of TPU and PDMS. The activity of mitochondrial enzymes and proliferation of human dermal fibroblasts significantly increased with the percentage of PDMS in the scaffolds [14].

Polyethylene terephthalate (PET) in combination with honey improved the morphology of chitosan-containing fibers, decreased the diameter of electrospun fibers, increased the fiber deposition area in the collector, and increased the water uptake capacities of the material, which is important for exudating wounds [15]. In another study, low-molecular weight cationic compounds were synthesized from re-purposed PET and used for self-assembling into high aspect ratio supramolecular nanofibers for encapsulation and delivery of piperacillin/tazobactam (PT), an anionic antibiotic. In a *Pseudomonas aeruginosa*-infected mouse skin wound model, the treatment with the PT-loaded nanofibers was more effective in comparison with free PT, as evidenced by significantly lower counts of *P. aeruginosa* at the wound sites, and by a histological analysis [43].

Polyethersulfone (PES) nanofibrous membranes, made by fine tuning of electrospinning parameters, supported the proliferation of fibroblasts similarly as standard tissue culture polystyrene, and when applied as experimental wound dressings in mice, they showed a higher exudate absorption capacity, higher epithelial regeneration, greater fibroblast maturation, improved collagen deposition, and faster edema resolution than control commercial wound dressings, namely Vaseline gauze dressing and a conventional gas permeable bandage [16].

Polystyrene (PS) in its amorphous state is a transparent and colorless material. It is a hard, stiff, and very brittle polymer with remarkable water vapor permeability, very high electrical resistance, and low dielectric loss. For wound dressing applications, PS was electrospun with poly(ε-caprolactone) and chamomile extract, containing phenolics and flavonoids, particularly apigenin with remarkable wound healing effect [44]. Electrospun polystyrene nanofibrous scaffolds were also applied for cultivation of skin cells in dynamic bioreactors and at the air/liquid interface [45, 46].

Poly(acrylic acid) (PAA) was recently used for preparation of nanofibers incorporated with reduced graphene oxide, intended for delivery of antibiotics, which was controlled by photothermal activation of the nanofibers [18]. In another study, electrospun nanofibers consisting of PAA and poly(1,8-octanediol-*co*-citric acid), i.e., a synthetic biodegradable elastomer, showed intrinsic antibacterial activity and were used for topical delivery of physiologically relevant concentrations of growth factors [47].

Poly(methyl methacrylate) (PMMA) nanofibers were used for construction of antiscarring wound dressings. PMMA containing polyethylene glycol and kynurenic acid, an antifibrotic agent, suppressed proliferation of fibroblasts *in vitro*, and when administered as wound dressing in rats *in vivo*, they inhibited expression of collagen and fibronectin, and enhanced the production of matrix metalloprotease 1 (MMP-1), an ECM-degrading enzyme [48]. In addition, core-shell nanofibers containing PVA and PMMA were used for delivery of ciprofloxacin hydrochloride, an antibiotic [49].

Poly(di(ethylene glycol) methyl ether methacrylate) (PDEGMA), a thermoresponsive polymer, was blended with poly(L-lactic acid-*co*-ε-caprolactone), P(LLA-CL), and used for construction of nanofibrous carriers for controlled drug delivery, namely for delivery of ciprofloxacin. These fibers also supported the growth of fibroblasts, and by decreasing the temperature, they enabled the cell detachment and delivery into wounds [50].

Another thermoresponsive polymer, poly(N-isopropylacrylamide) (PNIPAM) was used for fabrication of nanofibers for transdermal delivery of drugs, namely levothyroxine ($T_4$), which helps to reduce deposits of adipose tissue [51].

## 3. Nanofibers from synthetic degradable polymers

Synthetic degradable polymers have been used as scaffolds for skin tissue engineering, but also as wound dressing releasing various bioactive molecules by a controllable manner. Degradable polymers typically used in these applications are aliphatic polyesters, such as poly-ε-caprolactone (PCL), polylactide (PLA), and poly(lactide-*co*-glycolide) (PLGA). These polymers were approved by the Food and Drug Administration of the United States of America (FDA) for many medical applications.

Poly-ε-caprolactone (PCL) has been used most frequently from the mentioned polymers. It is a semi-crystalline polymer with tunable mechanical properties, and has a good solubility in a variety of solvents, and hence it can be combined with variety of other polymers. In comparison to other polyesters, PCL is a slowly degrading polymer, which can be essential for specific applications [52], and products of its degradation are non-toxic in the nature [53]. The acidic products of polyester degradation can affect the healing processes after implantation and may lead to inflammation [54]. However, due to the slow degradation of PCL, this risk is significantly lower compared to PLA and PLGA, which degrade significantly faster [55]. Slower degradation of PCL in comparison with its copolymer with PLA (PLCL) was also confirmed in our study, where both polymers were exposed to enzymatic degradation using lipase and proteinase K enzymes [13] (**Figure 2**).

However, PCL is more hydrophobic than PLA and particularly PLGA, and thus it is less supportive for cell adhesion. Therefore, PCL was rarely electrospun alone, i.e., without other polymers and bioactive additives. Nevertheless, pure PCL nanofibrous scaffolds were successfully used for cultivation and differentiation of hair follicle stem cells, isolated from the bulge regions of rat whiskers [56, 57]. In addition, pure PCL scaffolds supported the proliferation of mesenchymal stem cells, fibroblasts, and keratinocytes better than pure PVA scaffolds [12]. In spite of this, for purposes of skin regenerative therapies, PCL was usually combined with natural polymers, such as collagen, which was either blended with PCL before electrospinning [6, 58], or deposited on PCL nanofibers [59]. Gelatin, a collagen-derived protein, was either blended with PCL [60], or incorporated into core-shell PCL/gelatin nanofibers as the core polymer [22]. Gelatin was also electrospun independently of PCL using a double-nozzle technique, which resulted in creation of two types of nanofibers in the scaffolds, either mixed [61] or arranged in separate gelatin and PCL layers [27]. Multilayered and blend structures were found to fit most of native skin requirements in comparison with all the other mentioned structures [27].

Other natural polymers for modification of PCL nanofibers included whey protein [62], hyaluronic acid [24], keratin [28, 63], chitosan [28], fibrinogen [64], or gum arabic, and a corn protein zein [65]. These natural polymers were blended with PCL in the electrospinning solution. Polymer-modified PCL nanofibrous scaffolds have been often further modified with growth factors, such as epidermal growth factor (EGF) immobilized on PCL/collagen nanofibers [58] or on PCL/gelatin nanofibers [60] and transforming growth factor-β1 (TGF-β1) added into PCL/collagen electrospinning solution [6].

Other bioactive substances used for incorporation into PCL-based nanofibers included medicinal herbs such as *Aloe vera* [61], lawsone, i.e., 2-hydroxy-1,4-naphthoquinone extracted from Henna, endowed with antimicrobial, antiparasitic,

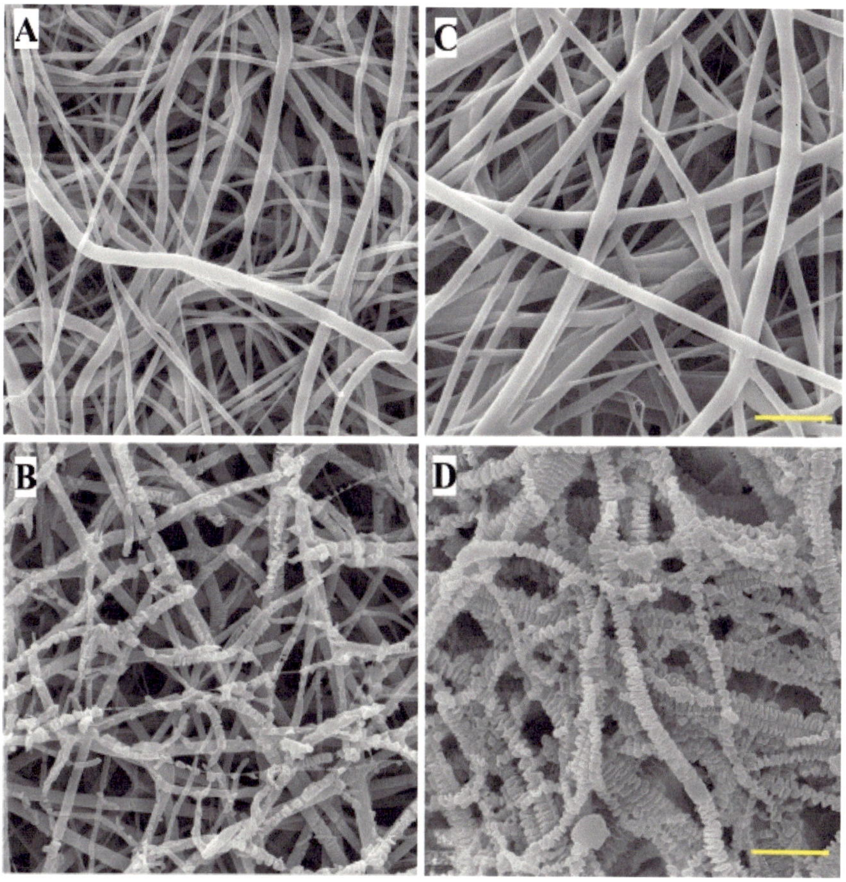


**Figure 2.**
*Scanning electron microscopy analyses of electrospun PCL (A, B) and PLCL (C, D) in their intact state (upper row, A, C) and after 2 days of enzymatic degradation (lower row, B, D). Magnification 5000×, scale bar 10 μm.*

anticancer, and antioxidant activities [22], other plant extracts with wound healing effects, e.g., from *Calendula officinalis* [65] or *Indigofera aspalathoides, Azadirachta indica, Memecylon edule, and Myristica andamanica* [66], molybdenum oxide nanoparticles for treating skin cancer [67] or antibiotics, which can be combined with PCL by various manners, e.g., through whey protein [62] or through micelles coating PCL/collagen nanofibers [6].

In our experiments, PCL electrospun nanofibrous membranes were impregnated with alaptide or L-arginine. Alaptide is a spirocyclic dipeptide, which was designed as an analogue of melanocyte-stimulating hormone release-inhibiting factor (MIF) and synthesized by Kasafirek et al. at the Research Institute for Pharmacy and Biochemistry in Prague, Czechoslovakia, in the 1980s of the twentieth century [68]. Alaptide showed a great potential for regeneration of the injured skin and also for enhanced transdermal penetration of drugs [69, 70]. Arginine is amino acid which is a precursor of nitric oxide, implicated in wound healing. Arginine promoted re-epithelization and vascularization of wounds [71] and supported proliferative, antiapoptotic, and immune defense functions of fibroblasts [72]. In our study, concentrations up to 2.5 wt.% of alaptide and up to 10 wt.% of L-arginine were used for fabrication of the membranes. Normal human dermal fibroblasts (NHDF) were cultivated on the membranes for 7 days. Alaptide-containing membranes were fully colonized by the cells up to the highest alaptide concentration (2.5 wt.%). However, the highest arginine concentration (10 wt.%) appeared as cytotoxic (**Figure 3**). One of the possible explanations is a dual effect of nitric oxide, which can act either as antioxidant or as oxidative agent [73].

Polylactide (PLA) is a polymer obtained by the ring-opening polymerization of lactide, i.e., cyclic dimer of lactic acid, as the monomer. The lactide has two enantiomers, namely L-lactide and D-lactide. Polymerization of each enantiomer alone results in creation of poly-L-lactide (PLLA) or poly-D-lactide (PDLA). Polymerization of a racemic mixture of L-lactide and D-lactide, i.e., mixture containing equal amounts of both enantiomers, gives rise of poly DL-lactide (PDLLA).

For skin tissue engineering, similarly as in PCL, PLA was often combined with other polymers and biologically active molecules in order to tailor desirable properties of the scaffolds. For example, for enhancing the cell adhesion on nanofibrous PLA scaffolds, PLA was blended and electrospun together with gelatin. Composite scaffolds containing PLA and gelatin in a ratio of 7:3 were more suitable for the attachment and viability of fibroblasts than the scaffolds made either of PLA or of gelatin alone [9]. Similarly, composite nanofibrous scaffolds made by electrospinning of a blend of poly-L-lactic acid/poly-(α,β)-DL-aspartic acid/collagen (PLLA/PAA/Col I&III) increased the proliferation of adipose tissue-derived stem cells (ADSCs), i.e., an important cell type used in skin tissue engineering, in comparison with pure PLLA or PLLA/PAA scaffolds [74]. In our experiments, electrospun PLLA meshes were modified by additional coating with fibrin or collagen. Fibrin coating supported better the growth of dermal fibroblasts, while the growth of keratinocytes was better on collagen [10].

In another design of bilayer scaffolds for skin tissue engineering, PLLA in the form of microporous disc was combined with superficial chitosan/PCL nanofibrous mat. The disc was seeded with dermal fibroblasts, while the mat was used as

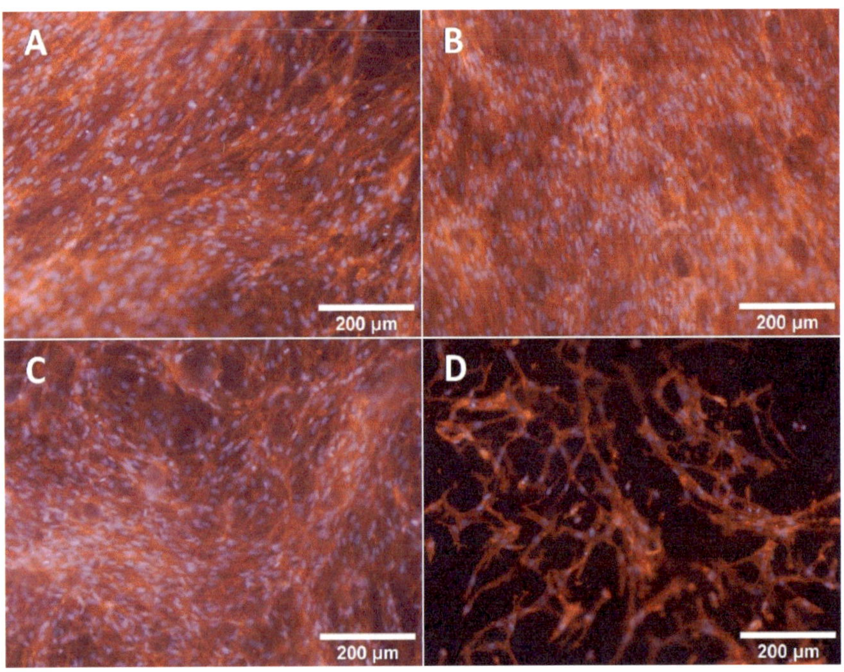


**Figure 3.**
*Normal human dermal fibroblasts cultivated for 7 days on PCL nanofibrous membrane impregnated with alaptide or arginine. A—0.1 wt.% of alaptide, B—2.5 wt.% of alaptide, C—1 wt.% of arginine, 10 wt.% of arginine. The cells were stained for nuclei (blue) and actin (red) using DNA-binding dye DAPI and phalloidin conjugated with TRITC. The images were acquired using Olympus IX71 fluorescence microscope equipped with DP71 camera and lens 10× (N.A. = 0.3).*

substrate for keratinocytes. The porous structure of the scaffolds allowed humoral communication of both cell types, but the nanofibers prevented the direct intermingling of these cell types [29].

Other interesting application of PLA nanofibers was skin tissue engineering for the infected wound site. PLA solution was electrospun together with highly porous silver microparticles (AgMPs) or high surface area silver nanoparticles (AgNPs) and used as substrates for co-culture of human epidermal keratinocytes and *Staphylococcus aureus*. The scaffolds with AgMPs showed a higher and steadier release of silver ions and lower cytotoxicity towards keratinocytes than AgNPs-loaded scaffolds [75].

PLA nanofibers have also been widely used for wound dressing applications, where they were loaded with various bioactive molecules improving wound healing and preventing microbial infection. Examples include PLLA/zein nanofibrous mats loaded with *Rana chensinensis* skin peptides with antibacterial and antioxidative activity [76], electrospun PLLA nanofibrous membranes coated by an *Aloe vera* gel [77], nanofibrillar matrices prepared from blends of PCL and PDLLA loaded with ciprofloxacin [78] or composite electrospun membranes containing polylactide:poly(vinyl pyrrolidone)/polylactide:poly(ethylene glycol) (PLA:PVP/PLA:PEG) core/shell fibers, designed for treatment of burns and loaded with curcumin and HHC36 antimicrobial peptides [8].

In spite of all these encouraging results, PLA and PCL can elicit inflammatory response. Although inflammation is the first physiological stage of wound healing, followed by proliferation and remodeling, excessive inflammation can delay the wound healing and can lead to ulceration, fibrosis, scar formation or entering the wound into a chronic state [79, 80]. The inflammatory response to PLA and PCL was reduced in electrospun co-axial scaffolds containing nanofibers with bioactive gelatin shells and biodegradable synthetic cores of PLA and PCL [81]. Another approach was the incorporation of PLA scaffolds with anti-inflammatory drugs. PLA nanofibers with 20 wt.% of ibuprofen promoted the viability and proliferation of human epidermal keratinocytes (HEK) and human dermal fibroblasts (HDF) *in vitro*, reduced wound contraction in mice *in vivo*, and when seeded with HEK and HDF, also enhanced new blood vessel formation in wounds of nude mice [80]. In a study by Yaru et al. [82], PLA nanofibers were incorporated with salicylate, a signaling molecule in plants, but also exhibiting a wide spectrum of signaling activities in mammals, including antithrombotic, anti-inflammatory, antineoplastic, and antimicrobial actions [83]. In addition, electrospun nanofibrous PDLLA scaffolds were incorporated with microalga *Spirulina*, which has anti-inflammatory, antioxidant, antimicrobial, antiallergenic, anticancer, and antidiabetic effects. The scaffolds were seeded with mesenchymal stem cells derived from mouse kidneys and used for treatment of the third degree burns in mice [79].

PLA and PCL can be combined in a poly(L-lactic acid-*co*-ε-caprolactone) copolymer, P(LLA-CL), also referred to as PLACL [84–86], PLLCL [26] or PLCL [13, 50].

As mentioned above, blend nanofibers of P(LLA-CL) and PDEGMA were prepared for controlled drug and cell delivery [50]. P(LLA-CL) was also blended with gelatin [84], silk fibroin, vitamin E, and curcumin [85] or with silk fibroin, tetracycline, and ascorbic acid [86], which increased the proliferation of human dermal fibroblasts on these nanofibrous scaffolds and secretion of collagen by these cells. Co-axial nanofibers with P(LLA-CL)/gelatin shell and albumin core containing EGF, insulin, hydrocortisone, and retinoic acid supported proliferation and epidermal differentiation of ADSCs better than nanofibers prepared by a blend spinning of all mentioned components [26]. In combination with poloxamer (Pluronic) 123,

P(LLA-CL) was also used for electrospinning of nanofibrous scaffolds for direct delivery of ADSCs into wounds in order to promote their healing [87].

In our experiments, composite PCL/PLCL nanofibers were coated either with platelet lysate, or with platelet lysate incorporated in fibrin assemblies [88] in various concentrations. Results for human keratinocytes (HaCaT cells) indicated that the presence of platelet lysate increased the metabolic activity and phenotypic maturation of keratinocytes. The best results were observed when the nanofibers were coated with fibrin together with platelet lysate (**Figure 4**).

Poly(lactide-*co*-glycolide) (PLGA) is a copolymer obtained by the ring-opening co-polymerization of two different monomers, i.e., lactic acid and glycolic acid. In skin regenerative therapies, it was applied for both skin tissue engineering and wound dressing. For these applications, PLGA was combined with various natural and synthetic polymers and bioactive compounds. For example, using bovine serum albumin as a carrier protein, vitamin C, vitamin D3, hydrocortisone, insulin, triiodothyronine, and EGF were simultaneously blend-spun into PLGA-collagen nanofibers. All these factors concertedly increased proliferation of fibroblasts and keratinocytes, while maintaining the keratinocyte basal state. In addition, vitamin C maintained its ability to facilitate secretion of type I collagen by fibroblasts, EGF stimulated proliferation of skin fibroblasts, and insulin potentiated adipogenic differentiation of fibroblasts [11]. In PLGA nanofibers, EGF was also combined with the local anesthetic lidocaine in order to accelerate wound healing in a rat model [89]. Coating PLGA nanofibers with a self-assembled complex of poly(ethylene argininyl aspartate diglyceride) polycation, heparin, and cargo growth factors, i.e., vascular endothelial growth factor (VEGF) and/or transforming growth factor-beta3 (TGF-β3), enhanced proliferation of human dermal fibroblasts and formation of tubular structures from human umbilical vein endothelial cells *in vitro*. In addition, these nanofibers reduced necrosis, improved vascularization, and maintained well-composed skin appendages in a mouse skin flap model *in vivo* [25]. Growth factors, namely recombinant human EGF and recombinant human basic fibroblast growth factor (bFGF), were also encapsulated in PLGA microspheres and loaded into hybrid scaffolds of PLGA and polyethylene oxide [90]. Both growth factors had a synergistic effect on the proliferation of human skin fibroblasts and increased the expression of genes for collagen and elastin in these cells [90]. Composite nanofibrous membranes containing PLGA and cellulose nanocrystals and loaded with neurotensin accelerated healing of full-thickness skin wounds in spontaneously diabetic mice [91]. Nanofibers created by electrospinning the dispersion composed of polyethyleneimine-carboxymethyl chitosan/pDNA-angiogenin

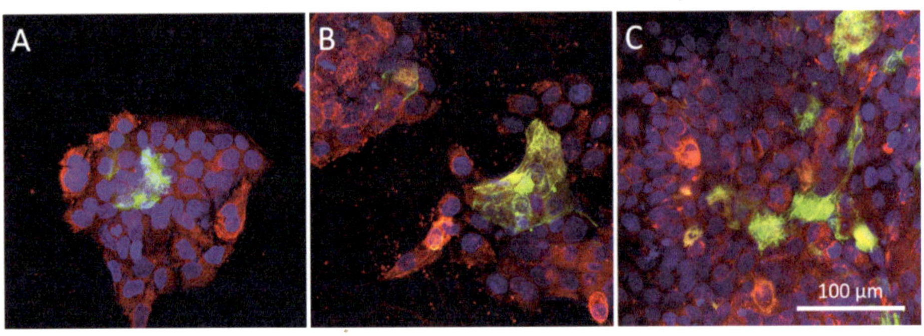


**Figure 4.**
*Immunofluorescence staining of cytokeratin 10 (green), cytokeratin 14 (red), and nuclei (blue) of HaCaT cells after 7 days in culture grown on PCL/PLCL nanofibers. Cell on nanofibers without coating (A), coated with platelet lysate (B) and coated with fibrin assemblies with platelet lysate (C) are shown. Leica TCS SPE DM2500 confocal microscope.*

nanoparticles, curcumin, PLGA, and cellulose nanocrystals showed antimicrobial and regenerative effects when transplanted into the infected full-thickness burn wounds in rats [92].

The PLGA nanofibers were also modified with ECM components. In a study by Shtrichman et al. [93], the PLGA nanofibrous scaffolds were modified with ECM deposited on these scaffolds by mesenchymal progenitor cells, derived from human embryonic stem cells, and human induced pluripotent stem cells, originating from hair follicle keratinocytes, which were cultured on the scaffolds and removed by subsequent decellularization. Subcutaneous implantation of the ECM-modified scaffolds in rats then showed that this stem cell-derived construct is biocompatible, biodegradable, and holds great potential for tissue regeneration applications. In addition, ECM-derived proteins, such as collagen and gelatin, can be electrospun directly together with PLGA [94]. In our earlier study, PLGA nanofibers were modified with fibrin or collagen in a similar manner as PLLA [10]. The morphology of these coatings, and also the behavior of HaCaT keratinocytes and human dermal fibroblasts on the coated and uncoated nanofibers, were similar on PLGA and PLLA [10].

PLGA-based nanofibrous meshes were also used for treatment of skin fibrosis and keloids, formed by abnormal proliferation of scar tissue at the site of cutaneous injury. Composite nanofibers of PLGA and poly(vinyl alcohol) loaded with kynurenine, a tryptophan metabolite, improved the dermal fibrosis in a rat model [95]. PLGA nanofibers releasing dexamethasone and green tea polyphenols significantly induced the degradation of collagen fibers in keloids on the back of nude mice [96].

Last but not least, PLGA nanofibers were also explored for transdermal delivery of drugs with poor oral absorption and limited bioavailability, e.g., Daidzein, a promising candidate for treating cardiovascular and cerebrovascular diseases [97], or for local delivery of anticancer drugs (for a review, see [98]).

Another important degradable polymer for fabrication of nanofibrous meshes for skin regenerative therapies is poly(ethylene glycol) (PEG), also known as poly(ethylene oxide) (PEO), depending on its molecular weight. PEG usually refers to polymers with a molecular mass below 20,000 g/mol, while PEO refers to polymers with a molecular mass above 20,000 g/mol.

In nanofibrous scaffolds, PEG or PEO have been usually used as auxiliary components improving electrospinnability, mechanical properties, and wettability of other polymers. For example, PEO was used to enable electrospinning of casein (i.e., a protein extensively used for drug delivery), which does not possess sufficient viscoelasticity due to its extensive intermolecular interactions [99], or to improve the electrospinnability and mechanical properties of silk fibroin [100]. As mentioned above, PEO or PEG was electrospun together with PMMA for creation of nanofibers delivering kynurenic acid [95] or with PLGA for delivery of human recombinant EGF and bFGF [90]. Other interesting applications of PEO include creation of electrospun carboxymethylcellulose/PEO nanofibers for delivery of viable commensal bacteria for preventive diabetic foot treatment [101], creation of three-dimensional scaffolds composed of PCL-PEG-PCL tri-block copolymer and iron oxide ($Fe_3O_4$) nanoparticles for skin tissue engineering [102], creation of biodegradable nanofiber mats based on thermoresponsive multiblock poly(ester urethane)s comprising PEG, poly(propylene glycol) (PPG), and PCL, which showed improved hydrolytic degradation compared to pure PCL and excellent adhesion of human dermal fibroblasts [103]. The adhesion and growth of fibroblast were also improved after combination of PLCL with Pluronic, i.e., a copolymer of PEO and poly(propylene oxide) (PPO) arranged in a tri-block PEO-PPO-PEO structure [104].

Other auxiliary polymers used for creation of nanofibrous scaffolds are poly(vinyl alcohol) (PVA) and poly(vinyl pyrrolidone) (PVP). In some studies,

PVA is regarded as hydrolytically degradable [17, 105], while in other studies, it is considered non-degradable [106]. PVP has been reported to be hydrolytically degradable [105]. In addition, both PVA and PVP are hydrophilic and water soluble, and thus they can be removed from a composite polymeric mesh in water environment. This property of PVA, PVP, and also of PEG or PEO, can be used for creation of so-called "sacrificial fibers" in order to enlarge the pores in nanofibrous scaffolds for penetration of cells [107]; for a review, see [108, 109] or for tailoring the appropriate surface roughness of nanofibers inside the scaffolds. For example, PLLA was electrospun together with PVP in increasing concentrations, and after subsequent etching of PVP from the scaffolds in water environment, nano- and microfibers with increasing nanoscale surface roughness were obtained. Higher surface nanoroughness and porosity of PLLA fibers increased their hydrophilicity and their colonization with human dermal fibroblasts [32]. Other applications of PVA and PVP are similar to those of PEG or PEO, i.e., to increase spinnability of poorly spinnable substances used for skin regenerative therapies. For this purpose, PVA was combined with polysaccharides, such as gum tragacanth [110] or Schizophyllan [111], and PVP with *Aloe vera* [112]. Both PVA and PVP have been used to improve mechanical properties, wettability, and attractiveness for cell adhesion of various synthetic and natural polymers, particularly PCL [110] and chitosan [1]. PVA was used as emulsifier in fabrication of blended electrospun PLGA/chitosan nanofibers for potential skin reconstruction [113]. PVA and particularly PVP are important components of nanofibers delivering various biomolecules and drugs into skin, such as antibiotics (PVA [49], PVP [114]), kynurenine (PVA [48]), curcumin and HHC36 antimicrobial peptides (PVP [8]) or antimicrobial suberin fatty acids isolated from outer birch bark (PVP [115]). PVA and PVP were combined in electrospun nanofibrous membranes designed for sustained release of the antibiotic ciprofloxacin into wounds [116]. Nanofibers with cellulose acetate (CA) as the core material and PVP solution as the shell material were used for transdermal delivery of artemisinin, a potent antimalarial drug, which was incorporated into CA [117].

## 4. Advanced skin tissue engineering

### 4.1 Dynamic bioreactors

Tissue engineering in general, including skin tissue engineering, can be markedly improved by cultivation of cell-material constructs in dynamic bioreactors. These systems not only improve the supply of oxygen and nutrients to cells and waste removal, but also mechanically stimulate the cells with positive effects on their growth, differentiation, and phenotypic maturation.

First of all, the cell seeding can be improved in dynamic systems. In a study by Vitacolonna et al. [118], various methods of seeding fibroblasts on acellular dermal matrix were compared, namely static cell seeding after previous degassing of the matrix using a low-pressure syringe system, orbital shaker seeding, centrifugal seeding, and their combinations. Centrifugal seeding combined with matrix degassing significantly increased the seeding efficiency and homogeneity compared to the other methods.

Also the subsequent proliferation and other performance of cells can be markedly influenced by the dynamic cultivation. For example, human epidermal stem cells cultured on microcarriers in a rotary bioreactor exhibited higher proliferation and viability than the cells cultured in static conditions [119]. Human fibroblasts on nanofibrous poly(3-hydroxybutyrate-co-3- hydroxyvalerate) (PHBV) scaffolds, subjected to biaxial distension for periods of time in a dynamic bioreactor, developed elastin fibers, whereas the cells on the same scaffolds cultured under static conditions showed negligible elastin production [120]. Cyclic uniaxial stretching of human HaCaT keratinocytes on collagen-silicon sheets induced the production of metalloproteinase 9 (MMP-9), a proteolytic enzyme necessary for keratinocyte migration, in these cells [121]. Strain also improved the mechanical strength of an engineered skin containing electrospun collagen scaffolds, human dermal fibroblasts, and epidermal keratinocytes, which was probably a result of enhanced epidermal cell proliferation, differentiation, and increased ECM production [34]. The keratinocyte differentiation under mechanical tension can be attributed to up-regulation of h2-calponin, which associates with actin stress fibers and decreases the cell proliferation rate (for a review, see [122]). Another type of mechanical stimulation implicated in keratinocyte differentiation is pressure stress, which increases the concentration of intracellular calcium, a stimulator of keratinocyte differentiation [35, 123].

In our experiments, we have developed a custom perfusion dynamic culture system allowing cell cultivation on elastic silicone membranes and generating cyclic pressure stress. First, these membranes were treated with plasma in order to increase their wettability and their ability to attach thin films made of nanofibrillar cellulose [124]. Afterwards, the porcine adipose tissue-derived stem cells were seeded on this surface. After 7 days of mechanical stimulation, a multilayered cell structure was observed in dynamic conditions, whereas in static conditions, only a cell monolayer was formed (**Figure 5**).

Increased concentration of calcium in keratinocytes and their differentiation can be also achieved by other means than mechanical stimulation, namely by stimulation with laser beam [125] or monodirectional pulsed electric current [126]. Electrical stimulation also enhanced the migration and proliferation of fibroblasts,

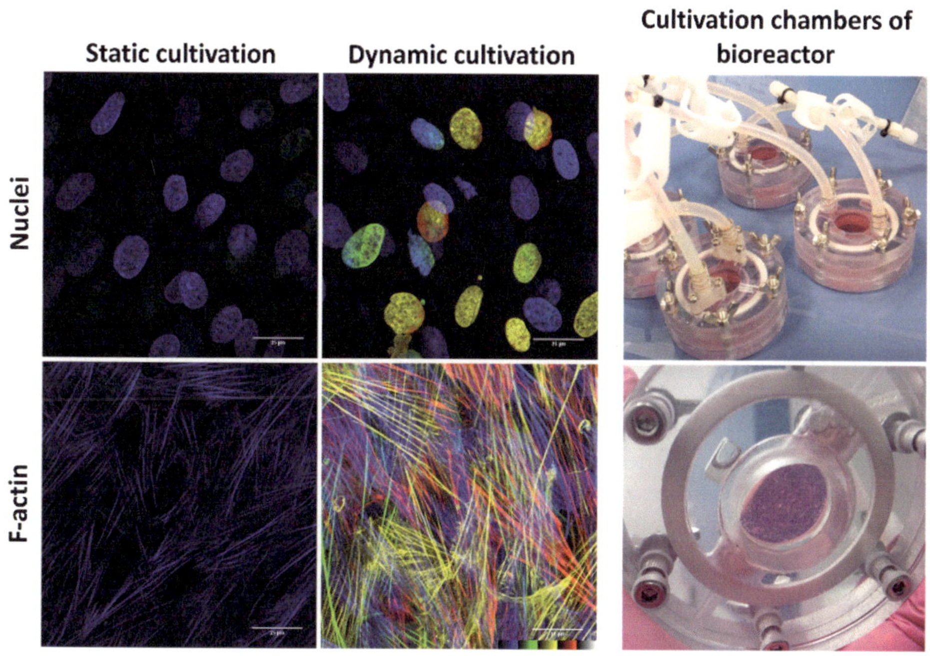


**Figure 5.**
*Color-coded projection of porcine adipose tissue-derived stem cells cultivated on thin nanocellulose film structure in static (left) and dynamic conditions (middle). Fluorescence staining of nuclei (DAPI) and F-actin (Phalloidin). Right: Custom built culture chambers creating controlled mechanical stress and strain with perfusion. Below, formation of a multilayered structure of cells creating opaque layer on the nanocellulose-coated silicone membrane.*

expression of ECM proteins in these cells, and differentiation of these cells towards myofibroblasts, i.e., processes critical for wound healing [127]. The positive effect of electrical current on fibroblasts can be further combined with light stimulation of the fibroblast proliferation, e.g., on nanofibrous PCL scaffolds electrospun with a semiconductive polymer, namely poly(N,N-bis(2-octyldodecyl)-3,6-di(thiophen-2-yl)-2,5-dihydropyrrolo[3,4-c]pyrrole-1,4-dione-alt-thieno[3,2-b]thiophene) (PDBTT), subjected to the illumination from a red light-emitting diode [128]. Also magnetic stimulation can be effectively used in skin tissue engineering. For example, multilayered sheets of keratinocytes were obtained by cultivation of keratinocytes loaded with magnetite cationic liposomes in a magnetic field. After removal of the magnet, the sheets were released from the cultivation plates, and were harvested with a magnet. This technology was termed "magnetic force-based tissue engineering" [129].

### 4.2 Air-liquid interface

In most experimental studies dealing with skin tissue engineering *in vitro*, keratinocytes are submerged in the cell culture media. However, in physiological skin *in vivo*, keratinocytes are exposed to air, at least their uppermost layer, i.e., *stratum corneum*, which is an impermeable barrier of cornified cell layers. Therefore, in advanced tissue engineering, keratinocytes should be exposed to the air-liquid interface (**Figure 6**) in order to achieve their phenotypic maturation and creation of the *stratum corneum* and the other epidermal layers, namely the basal, spinous, and granular layers [130].

The early differentiation of human amnion epithelial cells towards keratinocytes, manifested by formation desmosomes, was more pronounced in cells cultured at the air-liquid interface than in cells submerged in the culture medium [131]. In another study, human ADSCs were transdifferentiated towards keratinocytes in a medium containing retinoic acid, hydrocortisone, ascorbic acid, and bone morphogenetic protein-4 (BMP-4). This medium enabled high expression of pan-cytokeratin in conventional 2D cultures, especially if the cells were grown on type IV collagen. When the cell cultures were lifted to air-liquid interface, significant stratification was observed, particularly on growth supports coated with type IV collagen or fibronectin, and epidermal differentiation markers, such as involucrin and cytokeratins 1 and 14, were induced [132].

At the air-liquid interface, the keratinocytes or cells differentiating towards keratinocytes have been cultured on various substrates, e.g., acellular dermis [133], porous sponge-like gelatin scaffolds incorporated with chrondroitin-6-sulfate and hyaluronic acid [134], de-epithelialized human amniotic membrane [135], collagen IV and fibronectin [132] and fibrin in the form of layer [131], hydrogel or clot [136]. On the mentioned substrates, keratinocytes were grown either alone or in combination with fibroblasts submerged in the culture medium. In a study by Wang et al. [134], fibroblasts were grown inside the porous gelatin-based scaffolds submerged in the medium, while keratinocytes were grown on the top of the scaffolds, exposed to the air-liquid interface. Similarly, on the de-epithelialized amniotic membrane,

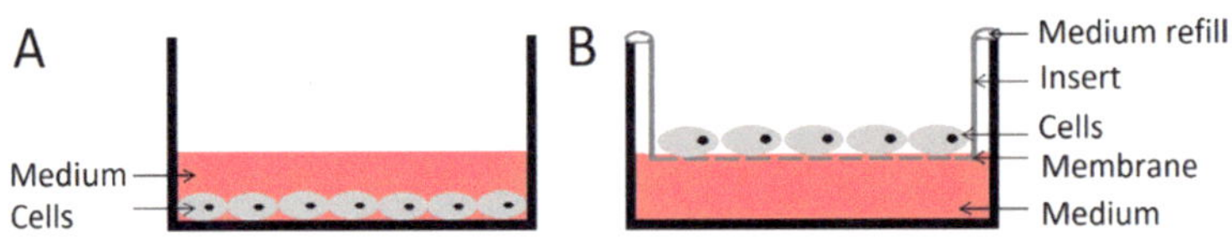


**Figure 6.**
*The principle of cell cultivation in a conventional cell culture system (A) and at the air-liquid interface (B).*

fibroblasts were cultured on the lower side of the membrane, submerged in the culture medium, while keratinocytes were grown on the upper side at the air-liquid interface [135]. In a study by Keck et al. [136], even a three-layered skin substitute was created. For the hypodermal layer, ADSCs and mature adipocytes were seeded within a fibrin hydrogel. On this layer, a fibrin clot with incorporated fibroblasts was placed for construction of the dermal layer. Keratinocytes were then added on the top of the two-layered construct and cultured at the air-liquid interface in order to create the epidermal layer [136].

Regarding the use of nanofibrous scaffolds for cultivation of keratinocytes at the air-liquid interface, synthetic and nature-derived scaffolds were used, namely electrospun PCL scaffolds [137, 138], electrospun polystyrene scaffolds [45] and fibrous sheets obtained after culturing human fibroblasts with ascorbic acid [139].

PCL scaffolds were used for construction of a three-dimensional *in vitro* skin model. The scaffolds were seeded with keratinocytes and melanocytes isolated from human scalp skin and cultured at the air-liquid interface. The keratinocytes contained a number of keratin fibrils and membrane-coated granules and formed a multilayered concentric structure, the surface of which became distinctly keratinized at the air-liquid interface. Cells with characteristic of melanocytes showed scattered distribution within the construct [137]. PCL scaffolds loaded with wound healing drugs, namely dexpanthenol and metyrapone, were used for a cell-based wound healing assay for rapid and predictive evaluation of wound therapeutics *in vitro*, using human HaCaT keratinocytes cultured at the air-liquid interface [138].

Interesting results were obtained on electrospun polystyrene scaffolds. In the absence of serum, keratinocytes, fibroblasts, and endothelial cells did not grow when cultured alone. However, when fibroblasts were cocultured with keratinocytes and endothelial cells, expansion of keratinocytes and endothelial cells occurred even in the absence of serum. Furthermore, the cells displayed native spatial three-dimensional organization when cultured at the air-liquid interface, even when all three cell types were introduced at random to the scaffolds [45].

The fibrous sheets produced by fibroblasts were used for creation of reconstructed human skin *in vitro*. After seeding the sheets with keratinocytes and the cell maturation *in vitro*, the reconstructed skin exhibited a well-developed human epidermis that expressed differentiation markers and basement membrane proteins [139].

The cultivation of keratinocytes at the air-liquid interface was also combined with cultivation of these cells in dynamic bioreactors, which further improved their growth and phenotypic maturation. Uniaxial strain stress (deformation of the cultivation substrate by 5–20%) further enhanced proliferation and epidermal differentiation of keratinocytes cultured at the air-liquid interface on electrospun collagen scaffolds containing fibroblasts in comparison with keratinocytes on unstrained cell-material constructs [34].

Also the perfusion with cell culture media showed beneficial effects on tissue-engineered skin constructs at the air-liquid interface. In a perfusion system with various growth supports for cells, such as acellular human dermis, Azowipes, electrospun polystyrene, and an electrospun composite of polystyrene and poly-DL-lactide fibers, human dermal fibroblast and endothelial cells showed greater viability under submerged conditions than at the air-liquid interface, whereas keratinocytes favored cultivation at the air-liquid interface. In addition, the viability of keratinocytes and fibroblasts was higher under continuous perfusion than under batch-feed perfusion, and on electrospun scaffolds than on acellular dermis [46]. In a recent study, a reconstructed skin model *in vitro*, containing a collagen matrix incorporated with fibroblasts and keratinocytes cultured at the air-liquid interface, was exposed to a continuous flow of cultivation medium (from 1.25 to

7.5 ml/h) at its basal side. Histological examination confirmed the formation of a significantly thicker *stratum corneum* compared to the control constructs cultivated under static conditions. Moreover, the keratinocyte differentiation markers involucrin and filaggrin, as well as the tight junction proteins claudin 1 and occludin, showed increased expression in the dynamically cultured skin models. However, the skin barrier function of the dynamically cultivated skin models was not enhanced compared with the skin models cultivated under static conditions [36]. Similar results were obtained in a study by Kalyanaraman et al. [140], performed on engineered skin substitutes based on collagen-glycosaminoglycan sponges, containing fibroblasts in their inside and keratinocytes on their surface, which were exposed to the air-liquid interface. Perfusion of these construct with the medium at the flow rate of 5 ml/min increased the metabolic activity of fibroblasts and maintained the epidermal barrier created by keratinocytes similarly as in static controls, while higher flow rates of 15 ml/min, and particularly 50 ml/min, decreased the cell metabolic activity, increased the degradation of the scaffolds and decreased the epidermal barrier function, manifested by its increased hydration [140].

## 5. Conclusions

Nanofibrous scaffolds made of synthetic polymers have been widely investigated for their potential use in skin regenerative therapies. Non-degradable polymers used for preparation of nanofibrous scaffolds included polyurethane (which can also be prepared in degradable form), polydimethylsiloxane (PDMS), polyethylene terephthalate (PET), polyethersulfone (PES), and even polystyrene (PS). These scaffolds were mainly intended for wound dressing applications, and in case of PS, also for cultivation of skin cells in dynamic bioreactor and at the air/liquid interface. For creation of nanofibrous meshes, the non-degradable polymers have been often used in combinations with nature-derived polymers (dextran, chitosan, gelatin, and keratin), and loaded with various wound healing, angiogenic, antioxidant, anti-inflammatory, photoprotective, and antimicrobial substances. Non-degradable synthetic polymers also include hydrogels, such as poly(acrylic acid) (PAA), poly(methyl methacrylate) (PMMA) and particularly poly(di(ethylene glycol) methyl ether methacrylate) (PDEGMA), which is thermoresponsive and suitable for controlled drug delivery and cell delivery into wounds. Degradable synthetic polymers have been also applied in wound healing, but also as direct scaffolds for skin tissue engineering, i.e., as carriers for keratinocytes, fibroblasts, and stem cells. The most widely used degradable polymers for these applications include polycaprolactone (PCL) and its copolymers with polylactides (PLCL), and also polylactides (PLLA and PDLLA) and their copolymers with polyglycolides (PLGA). Similarly as non-degradable polymers, also degradable polymers are almost exclusively used in combination with nature-derived polymers (collagen, gelatin, keratin, fibrin, and glycosaminoglycans) in order to increase their attractiveness for cell colonization, and also with some synthetic polymers, such as poly(ethylene glycol) (PEG), poly(ethylene oxide) (PEO), poly (vinyl alcohol) (PVA), and poly(vinyl pyrrolidone) (PVP). These synthetic polymers act as auxiliary, i.e., improving electrospinnability, mechanical properties, and wettability of other polymers. Similarly as non-degradable polymers, also degradable polymers have been loaded with a wide range of growth and angiogenic factors and other biologically active substances. The cell performance on non-degradable and degradable nanofibrous scaffolds can be further markedly improved by cultivation in dynamic bioreactors and/or at air/liquid interface.

## Acknowledgements

This review article was supported by the Grant Agency of the Czech Republic (grant No. 17-02448S) and the Ministry of Health of the Czech Republic (grant No. NV18-01-00332).

## Author details

Lucie Bacakova[1*], Marketa Zikmundova[1], Julia Pajorova[1], Antonin Broz[1], Elena Filova[1], Andreu Blanquer[1], Roman Matejka[1], Jana Stepanovska[1], Petr Mikes[2], Vera Jencova[2], Eva Kuzelova Kostakova[2] and Alla Sinica[3,4]

1 Department of Biomaterials and Tissue Engineering, Institute of Physiology of the Czech Academy of Sciences, Prague, Czech Republic

2 Faculty of Sciences, Humanities and Education, Technical University of Liberec, Liberec, Czech Republic

3 BIOCEV, First Faculty of Medicine, Charles University, Vestec, Czech Republic

4 Department of Analytical Chemistry, University of Chemistry and Technology Prague, Prague, Czech Republic

*Address all correspondence to: lucie.bacakova@fgu.cas.cz

## References

[1] Mahmoudi N, Eslahi N, Mehdipour A, Mohammadi M, Akbari M, Samadikuchaksaraei A, et al. Temporary skin grafts based on hybrid graphene oxide-natural biopolymer nanofibers as effective wound healing substitutes: Pre-clinical and pathological studies in animal models. Journal of Materials Science. Materials in Medicine. 2017;**28**(5):73. DOI: 10.1007/s10856-017-5874-y

[2] Bacakova L, Pajorova J, Bacakova M, Skogberg A, Kallio P, Kolarova K, et al. Versatile application of nanocellulose: From industry to skin tissue engineering and wound healing. Nanomaterials (Basel). 2019;**9**(2):164. DOI: 10.3390/nano9020164

[3] Bacakova L, Bacakova M, Pajorova J, Kudlackova R, Stankova L, Musilkova J, et al. Chapter 3: Nanofibrous scaffolds as promising cell carriers for tissue engineering. In: Rahman MM, Asiri AM, editors. ISBN 978-953-51-2529-7, Print ISBN 978-953-51-2528-0Nanofiber Research—Reaching New Heights. London, United Kingdom: IntechOpen; 2016. pp. 29-54. DOI: 10.5772/63707

[4] Manoukian OS, Matta R, Letendre J, Collins P, Mazzocca AD, Kumbar SG. Electrospun nanofiber scaffolds and their hydrogel composites for the engineering and regeneration of soft tissues. Methods in Molecular Biology. 2017;**1570**:261-278. DOI: 10.1007/978-1-4939-6840-4_18

[5] Rasouli R, Barhoum A, Bechelany M, Dufresne A. Nanofibers for biomedical and healthcare applications. Macromolecular Bioscience. 2019;**19**(2):e1800256. DOI: 10.1002/mabi.201800256

[6] Albright V, Xu M, Palanisamy A, Cheng J, Stack M, Zhang B, et al. Micelle-coated, hierarchically structured nanofibers with dual-release capability for accelerated wound healing and infection control. Advanced Healthcare Materials. 2018;**7**(11):e1800132. DOI: 10.1002/adhm.201800132

[7] Unnithan AR, Sasikala AR, Murugesan P, Gurusamy M, Wu D, Park CH, et al. Electrospun polyurethane-dextran nanofiber mats loaded with estradiol for post-menopausal wound dressing. International Journal of Biological Macromolecules. 2015;**77**:1-8. DOI: 10.1016/j.ijbiomac.2015.02.044

[8] Li W, Yu Q, Yao H, Zhu Y, Topham PD, Yue K, et al. Superhydrophobic hierarchical fiber/bead composite membranes for efficient treatment of burns. Acta Biomaterialia. 2019;**92**:60-70. DOI: 10.1016/j.actbio.2019.05.025

[9] Hoveizi E, Nabiuni M, Parivar K, Rajabi-Zeleti S, Tavakol S. Functionalisation and surface modification of electrospun polylactic acid scaffold for tissue engineering. Cell Biology International. 2014;**38**(1):41-49. DOI: 10.1002/cbin.10178

[10] Bacakova M, Pajorova J, Stranska D, Hadraba D, Lopot F, Riedel T, et al. Protein nanocoatings on synthetic polymeric nanofibrous membranes designed as carriers for skin cells. International Journal of Nanomedicine. 2017;**12**:1143-1160. DOI: 10.2147/IJN.S121299

[11] Peh P, Lim NS, Blocki A, Chee SM, Park HC, Liao S, et al. Simultaneous delivery of highly diverse bioactive compounds from blend electrospun fibers for skin wound healing. Bioconjugate Chemistry. 2015;**26**(7):1348-1358. DOI: 10.1021/acs.bioconjchem.5b00123

[12] Buzgo M, Plencner M, Rampichova M, Litvinec A, Prosecka E,

Staffa A, et al. Poly-ε-caprolactone and polyvinyl alcohol electrospun wound dressings: Adhesion properties and wound management of skin defects in rabbits. Regenerative Medicine. 2019;**14**(5):423-445. DOI: 10.2217/rme-2018-0072

[13] Mikes P, Horakova J, Saman A, Vejsadova L, Topham P, Punyodom W, et al. Comparison and characterization of different polyester nano/micro fibres for use in tissue engineering applications. Journal of Industrial Textiles. 2019. in press. DOI: 10.1177/1528083719848155

[14] Drupitha MP, Bankoti K, Pal P, Das B, Parameswar R, Dhara S, et al. Morphology-induced physico-mechanical and biological characteristics of TPU-PDMS blend scaffolds for skin tissue engineering applications. Journal of Biomedical Materials Research. Part B, Applied Biomaterials. 2019;**107**(5):1634-1644. DOI: 10.1002/jbm.b.34256

[15] Arslan A, Simşek M, Aldemir SD, Kazaroğlu NM, Gümüşderelioğlu M. Honey-based PET or PET/chitosan fibrous wound dressings: Effect of honey on electrospinning process. Journal of Biomaterials Science. Polymer Edition. 2014;**25**(10):999-1012. DOI: 10.1080/09205063.2014.918455

[16] Babaeijandaghi F,Shabani I, Seyedjafari E, Naraghi ZS, Vasei M, Haddadi-Asl V, et al. Accelerated epidermal regeneration and improved dermal reconstruction achieved by polyethersulfone nanofibers. Tissue Engineering. Part A. 2010;**16**(11):3527-3536. DOI: 10.1089/ten.TEA.2009.0829

[17] Gaaz TS, Sulong AB, Akhtar MN, Kadhum AA, Mohamad AB, Al-Amiery AA. Properties and applications of polyvinyl alcohol, halloysite nanotubes and their nanocomposites. Molecules. 2015;**20**(12):22833-22847. DOI: 10.3390/molecules201219884

[18] Altinbasak I, Jijie R, Barras A, Golba B, Sanyal R, Bouckaert J, et al. Reduced graphene-oxide-embedded polymeric nanofiber mats: An "on-demand" photothermally triggered antibiotic release platform. ACS Applied Materials & Interfaces. 2018;**10**(48):41098-41106. DOI: 10.1021/acsami.8b14784

[19] Amna T, Hassan MS, Yang J, Khil MS, Song KD, Oh JD, et al. Virgin olive oil blended polyurethane micro/nanofibers ornamented with copper oxide nanocrystals for biomedical applications. International Journal of Nanomedicine. 2014;**9**:891-898. DOI: 10.2147/IJN.S54113 (eCollection 2014)

[20] Hong SM, Kim JW, Knowles JC, Gong MS. Facile preparation of antibacterial, highly elastic silvered polyurethane nanofiber fabrics using silver carbamate and their dermal wound healing properties. Journal of Biomaterials Applications. 2017;**31**(7):1026-1038. DOI: 10.1177/0885328216687665

[21] Gomes S, Rodrigues G, Martins G, Henriques C, Silva JC. Evaluation of nanofibrous scaffolds obtained from blends of chitosan, gelatin and polycaprolactone for skin tissue engineering. International Journal of Biological Macromolecules. 2017;**102**:1174-1185. DOI: 10.1016/j.ijbiomac.2017.05.004

[22] Adeli-Sardou M, Yaghoobi MM, Torkzadeh-Mahani M, Dodel M. Controlled release of lawsone from polycaprolactone/gelatin electrospun nano fibers for skin tissue regeneration. International Journal of Biological Macromolecules. 2019;**124**:478-491. DOI: 10.1016/j.ijbiomac.2018.11.237

[23] Yao CH, Lee CY, Huang CH, Chen YS, Chen KY. Novel bilayer

wound dressing based on electrospun gelatin/keratin nanofibrous mats for skin wound repair. Materials Science & Engineering. C, Materials for Biological Applications. 2017;**79**:533-540. DOI: 10.1016/j.msec.2017.05.076

[24] Qian Y, Li L, Jiang C, Xu W, Lv Y, Zhong L, et al. The effect of hyaluronan on the motility of skin dermal fibroblasts in nanofibrous scaffolds. International Journal of Biological Macromolecules. 2015;**79**:133-143. DOI: 10.1016/j.ijbiomac.2015.04.059

[25] Lee MS, Ahmad T, Lee J, Awada HK, Wang Y, Kim K, et al. Dual delivery of growth factors with coacervate-coated poly(lactic-co-glycolic acid) nanofiber improves neovascularization in a mouse skin flap model. Biomaterials. 2017;**124**:65-77. DOI: 10.1016/j.biomaterials.2017.01.036

[26] Jin G, Prabhakaran MP, Kai D, Ramakrishna S. Controlled release of multiple epidermal induction factors through core-shell nanofibers for skin regeneration. European Journal of Pharmaceutics and Biopharmaceutics. 2013;**85**(3 Pt A):689-698. DOI: 10.1016/j.ejpb.2013.06.002

[27] Dias JR, Baptista-Silva S, Sousa A, Oliveira AL, Bártolo PJ, Granja PL. Biomechanical performance of hybrid electrospun structures for skin regeneration. Materials Science & Engineering. C, Materials for Biological Applications. 2018;**93**:816-827. DOI: 10.1016/j.msec.2018.08.050

[28] Zahedi E, Esmaeili A, Eslahi N, Shokrgozar MA, Simchi A. Fabrication and characterization of core-shell electrospun fibrous mats containing medicinal herbs for wound healing and skin tissue engineering. Marine Drugs. 2019;**17**(1):27. DOI: 10.3390/md17010027

[29] Lou T, Leung M, Wang X, Chang JY, Tsao CT, Sham JG, et al. Bi-layer scaffold of chitosan/PCL-nanofibrous mat and PLLA-microporous disc for skin tissue engineering. Journal of Biomedical Nanotechnology. 2014;**10**(6):1105-1113. DOI: 10.1166/jbn.2014.1793

[30] Pal P, Dadhich P, Srivas PK, Das B, Maulik D, Dhara S. Bilayered nanofibrous 3D hierarchy as skin rudiment by emulsion electrospinning for burn wound management. Biomaterials Science. 2017;**5**(9):1786-1799. DOI: 10.1039/c7bm00174f

[31] Bacakova M, Pajorova J, Broz A, Hadraba D, Lopot F, Zavadakova A, et al. A two-layer skin construct consisting of a collagen hydrogel reinforced by a fibrin-coated polylactide nanofibrous membrane. International Journal of Nanomedicine. 2019;**14**:5033-5050. DOI: 10.2147/IJN.S200782

[32] Ren L, Pandit V, Elkin J, Denman T, Cooper JA, Kotha SP. Large-scale and highly efficient synthesis of micro- and nano-fibers with controlled fiber morphology by centrifugal jet spinning for tissue regeneration. Nanoscale. 2013;**5**(6):2337-2345. DOI: 10.1039/c3nr33423f

[33] Loordhuswamy AM, Krishnaswamy VR, Korrapati PS, Thinakaran S, Rengaswami GD. Fabrication of highly aligned fibrous scaffolds for tissue regeneration by centrifugal spinning technology. Materials Science & Engineering. C, Materials for Biological Applications. 2014;**42**:799-807. DOI: 10.1016/j.msec.2014.06.011

[34] Powell HM, McFarland KL, Butler DL, Supp DM, Boyce ST. Uniaxial strain regulates morphogenesis, gene expression, and tissue strength in engineered skin. Tissue Engineering. Part A. 2010;**16**(3):1083-1092. DOI: 10.1089/ten.TEA.2009.0542

[35] Kobayashi Y, Sanno Y, Sakai A, Sawabu Y, Tsutsumi M, Goto M, et al.

Mathematical modeling of calcium waves induced by mechanical stimulation in keratinocytes. PLoS One. 2014;**9**(3):e92650. DOI: 10.1371/journal.pone.0092650

[36] Strüver K, Friess W, Hedtrich S. Development of a perfusion platform for dynamic cultivation of in vitro skin models. Skin Pharmacology and Physiology. 2017;**30**(4):180-189. DOI: 10.1159/000476071

[37] Yu K, Zhu T, Wu Y, Zhou X, Yang X, Wang J, et al. Incorporation of amoxicillin-loaded organic montmorillonite into poly(ester-urethane) urea nanofibers as a functional tissue engineering scaffold. Colloids and Surfaces. B, Biointerfaces. 2017;**151**:314-323. DOI: 10.1016/j.colsurfb.2016.12.034

[38] Kim JI, Pant HR, Sim HJ, Lee KM, Kim CS. Electrospun propolis/polyurethane composite nanofibers for biomedical applications. Materials Science & Engineering. C, Materials for Biological Applications. 2014;**44**:52-57. DOI: 10.1016/j.msec.2014.07.062

[39] Klempaiová M, Dragúňová J, Kabát P, Hnátová M, Koller J, Bakoš D. Cytotoxicity testing of a polyurethane nanofiber membrane modified with chitosan/β-cyclodextrin/berberine suitable for wound dressing application: Evaluation of biocompatibility. Cell and Tissue Banking. 2016;**17**(4):665-675. DOI: 10.1007/s10561-016-9585-2

[40] Maharjan B, Joshi MK, Tiwari AP, Park CH, Kim CS. In-situ synthesis of AgNPs in the natural/synthetic hybrid nanofibrous scaffolds: Fabrication, characterization and antimicrobial activities. Journal of the Mechanical Behavior of Biomedical Materials. 2017;**65**:66-76. DOI: 10.1016/j.jmbbm.2016.07.034

[41] Heo DN, Yang DH, Lee JB, Bae MS, Kim JH, Moon SH, et al. Burn-wound healing effect of gelatin/polyurethane nanofiber scaffold containing silver-sulfadiazine. Journal of Biomedical Nanotechnology. 2013;**9**(3):511-515. DOI: 10.1166/jbn.2013.1509

[42] Gencturk A, Kahraman E, Güngör S, Özhan G, Özsoy Y, Sarac AS. Polyurethane/hydroxypropyl cellulose electrospun nanofiber mats as potential transdermal drug delivery system: Characterization studies and in vitro assays. Artificial Cells, Nanomedicine, and Biotechnology. 2017;**45**(3):655-664. DOI: 10.3109/21691401.2016.1173047

[43] Liu S, Fukushima K, Venkataraman S, Hedrick JL, Yang YY. Supramolecular nanofibers self-assembled from cationic small molecules derived from repurposed poly(ethylene teraphthalate) for antibiotic delivery. Nanomedicine. 2018;**14**(1):165-172. DOI: 10.1016/j.nano.2017.09.007

[44] Motealleh B, Zahedi P, Rezaeian I, Moghimi M, Abdolghaffari AH, Zarandi MA. Morphology, drug release, antibacterial, cell proliferation, and histology studies of chamomile-loaded wound dressing mats based on electrospun nanofibrous poly(ε-caprolactone)/polystyrene blends. Journal of Biomedical Materials Research. Part B, Applied Biomaterials. 2014;**102**(5):977-987. DOI: 10.1002/jbm.b.33078

[45] Sun T, Mai S, Norton D, Haycock JW, Ryan AJ, MacNeil S. Self-organization of skin cells in three-dimensional electrospun polystyrene scaffolds. Tissue Engineering. 2005;**11**(7-8):1023-1033. DOI: 10.1089/ten.2005.11.1023

[46] Sun T, Norton D, Haycock JW, Ryan AJ, MacNeil S. Development of a closed bioreactor system for culture of tissue-engineered skin at an air-liquid interface. Tissue Engineering. 2005;**11**(11-12):1824-1831. DOI: 10.1089/ten.2005.11.1824

[47] Goins A, Ramaswamy V, Dirr E, Dulany K, Irby S, Webb A, et al. Development of poly (1,8 octanediol-co-citrate) and poly (acrylic acid) nanofibrous scaffolds for wound healing applications. Biomedical Materials. 2017;**13**(1):015002. DOI: 10.1088/1748-605X/aa8439

[48] Poormasjedi-Meibod MS, Pakyari M, Jackson JK, Salimi Elizei S, Ghahary A. Development of a nanofibrous wound dressing with an antifibrogenic properties in vitro and in vivo model. Journal of Biomedical Materials Research. Part A. 2016;**104**(9):2334-2344. DOI: 10.1002/jbm.a.35770

[49] Zupančič Š, Sinha-Ray S, Sinha-Ray S, Kristl J, Yarin AL. Controlled release of ciprofloxacin from core-shell nanofibers with monolithic or blended core. Molecular Pharmaceutics. 2016;**13**(4):1393-1404. DOI: 10.1021/acs.molpharmaceut.6b00039

[50] Li H, Williams GR, Wu J, Lv Y, Sun X, Wu H, et al. Thermosensitive nanofibers loaded with ciprofloxacin as antibacterial wound dressing materials. International Journal of Pharmaceutics. 2017;**517**(1-2):135-147. DOI: 10.1016/j.ijpharm.2016.12.008

[51] Azarbayjani AF, Venugopal JR, Ramakrishna S, Lim PF, Chan YW, Chan SY. Smart polymeric nanofibers for topical delivery of levothyroxine. Journal of Pharmacy & Pharmaceutical Sciences. 2010;**13**(3):400-410

[52] Ali Akbari Ghavimi S, Ebrahimzadeh MH, Solati-Hashjin M, Abu Osman NA. Polycaprolactone/starch composite: Fabrication, structure, properties, and applications. Journal of Biomedial Materials Research Part A. 2015;**103**:2482-2498. DOI: 10.1002/jbm.a.35371

[53] Wang Y, Rodriguez-Perez MA, Reis RL, Mano JF. Thermal and thermomechanical behaviour of polycaprolactone and starch/polycaprolactone blends for biomedical applications. Macromolecular Materials and Engineering. 2005;**290**:792-801. DOI: 10.1002/mame.20050 0003

[54] Ceonzo K, Gaynor A, Shafferm L, et al. Polyglycolic acid-induced inflammation: Role of hydrolysis and resulting complement activation. Tissue Engineering. 2006;**12**:301-308. DOI: 10.1089/ten.2006.12.301

[55] Woodruff MA, Hutmacher DW. The return of a forgotten polymer—Polycaprolactone in the 21st century. Progress in Polymer Science. 2010;**35**:1217-1256. DOI: 10.1016/j.progpolymsci.2010.04.002

[56] Hejazian LB, Esmaeilzade B, Moghanni Ghoroghi F, Moradi F, Hejazian MB, Aslani A, et al. The role of biodegradable engineered nanofiber scaffolds seeded with hair follicle stem cells for tissue engineering. Iranian Biomedical Journal. 2012;**16**(4):193-201. DOI: 10.6091/ibj.1074.2012

[57] Yari A, Teimourian S, Amidi F, Bakhtiyari M, Heidari F, Sajedi N, et al. The role of biodegradable engineered random polycaprolactone nanofiber scaffolds seeded with nestin-positive hair follicle stem cells for tissue engineering. Advanced Biomedical Research. 2016;**5**:22. DOI: 10.4103/2277-9175.175911

[58] Gümüşderelioğlu M, Dalkıranoğlu S, Aydın RS, Cakmak S. A novel dermal substitute based on biofunctionalized electrospun PCL nanofibrous matrix. Journal of Biomedical Materials Research. Part A. 2011;**98**(3):461-472. DOI: 10.1002/jbm.a.33143

[59] Ghosal K, Manakhov A, Zajíčková L, Thomas S. Structural and surface compatibility study of modified electrospun poly(ε-caprolactone) (PCL) composites for skin tissue engineering.

AAPS PharmSciTech. 2017;**18**(1):72-81. DOI: 10.1208/s12249-016-0500-8

[60] Tığlı RS, Kazaroğlu NM, Mavış B, Gümüşderelioğlu M. Cellular behavior on epidermal growth factor (EGF)-immobilized PCL/gelatin nanofibrous scaffolds. Journal of Biomaterials Science. Polymer Edition. 2011;**22**(1-3):207-223. DOI: 10.1163/092050609X12591500475424

[61] Baghersad S, Hajir Bahrami S, Mohammadi MR, Mojtahedi MRM, Milan PB. Development of biodegradable electrospun gelatin/aloe-vera/poly(ε-caprolactone) hybrid nanofibrous scaffold for application as skin substitutes. Materials Science & Engineering. C, Materials for Biological Applications. 2018;**93**:367-379. DOI: 10.1016/j.msec.2018.08.020

[62] Ahmed SM, Ahmed H, Tian C, Tu Q, Guo Y, Wang J. Whey protein concentrate doped electrospun poly(epsilon-caprolactone) fibers for antibiotic release improvement. Colloids and Surfaces. B, Biointerfaces. 2016;**143**:371-381. DOI: 10.1016/j.colsurfb.2016.03.059

[63] Cruz-Maya I, Guarino V, Almaguer-Flores A, Alvarez-Perez MA, Varesano A, Vineis C. Highly polydisperse keratin rich nanofibers: Scaffold design and in vitro characterization. Journal of Biomedical Materials Research. Part A. 2019;**107**(8):1803-1813. DOI: 10.1002/jbm.a.36699

[64] Mirzaei-Parsa MJ, Ghanbari H, Alipoor B, Tavakoli A, Najafabadi MRH, Faridi-Majidi R. Nanofiber-acellular dermal matrix as a bilayer scaffold containing mesenchymal stem cell for healing of full-thickness skin wounds. Cell and Tissue Research. 2019;**375**(3):709-721. DOI: 10.1007/s00441-018-2927-6

[65] Rad ZP, Mokhtari J, Abbasi M. Calendula officinalis extract/PCL/Zein/gum arabic nanofibrous bio-composite scaffolds via suspension, two-nozzle and multilayer electrospinning for skin tissue engineering. International Journal of Biological Macromolecules. 2019. DOI: 10.1016/j.ijbiomac.2019.05.204

[66] Jin G, Prabhakaran MP, Kai D, Annamalai SK, Arunachalam KD, Ramakrishna S. Tissue engineered plant extracts as nanofibrous wound dressing. Biomaterials. 2013b;**34**(3):724-734. DOI: 10.1016/j.biomaterials.2012.10.026

[67] Janani I, Lakra R, Kiran MS, Korrapati PS. Selectivity and sensitivity of molybdenum oxide-polycaprolactone nanofiber composites on skin cancer: Preliminary in-vitro and in-vivo implications. Journal of Trace Elements in Medicine and Biology. 2018;**49**:60-71. DOI: 10.1016/j.jtemb.2018.04.033

[68] Kasafírek E, Rybák M, Krejcí I, Sturc A, Krepela E, Sedo A. Two-step generation of spirocyclic dipeptides from linear peptide ethyl ester precursors. Life Sciences. 1992;**50**(3):187-193. DOI: 10.1016/0024-3205(92)90271-p

[69] Opatrilova R, Cernikova A, Coufalova L, Dohnal J, Jampilek J. In vitro permeation of micronized and nanonized alaptide from semisolid formulations. ScientificWorldJournal. 2013;**2013**:787283. DOI: 10.1155/2013/787283 eCollection 2013

[70] Pokorna A, Bobal P, Oravec M, Rarova L, Bobalova J, Jampilek J. Investigation of permeation of theophylline through skin using selected piperazine-2,5-Diones. Molecules. 2019;**24**(3):566. DOI: 10.3390/molecules24030566

[71] Feizabadi F, Minaiyan M, Taheri A. Arginine functionalized bacterial cellulose nanofibers containing gel as an effective wound dressing: In vitro

and in vivo evaluation. Current Drug Delivery. 2018;**15**(6):840-849. DOI: 10.2174/1567201815666180220100154

[72] Kocic H, Arsic I, Stankovic M, Tiodorovic D, Ciric V, Kocic G. Proliferative, anti-apoptotic and immune-enhancing effects of L-arginine in culture of skin fibroblasts. Journal of Biological Regulators and Homeostatic Agents. 2017;**31**(3):667-672

[73] Hampl V, Herget J. Role of nitric oxide in the pathogenesis of chronic pulmonary hypertension. Physiological Reviews. 2000;**80**(4):1337-1372. DOI: 10.1152/physrev.2000.80.4.1337

[74] Ravichandran R, Venugopal JR, Sundarrajan S, Mukherjee S, Sridhar R, Ramakrishna S. Composite poly-L-lactic acid/poly-(α,β)-DL-aspartic acid/collagen nanofibrous scaffolds for dermal tissue regeneration. Materials Science & Engineering. C, Materials for Biological Applications. 2012;**32**(6):1443-1451. DOI: 10.1016/j.msec.2012.04.024

[75] Mohiti-Asli M, Pourdeyhimi B, Loboa EG. Skin tissue engineering for the infected wound site: Biodegradable PLA nanofibers and a novel approach for silver ion release evaluated in a 3D coculture system of keratinocytes and Staphylococcus aureus. Tissue Engineering. Part C, Methods. 2014;**20**(10):790-797. DOI: 10.1089/ten.TEC.2013.0458

[76] Zhang M, Li X, Li S, Liu Y, Hao L. Electrospun poly(L-lactide)/zein nanofiber mats loaded with Rana chensinensis skin peptides for wound dressing. Journal of Materials Science: Materials in Medicine. 2016;**27**(9):136. DOI: 10.1007/s10856-016-5749-7

[77] Jouybar A, Seyedjafari E, Ardeshirylajimi A, Zandi-Karimi A, Feizi N, Khani MM, et al. Enhanced skin regeneration by herbal extract-coated poly-L-lactic acid nanofibrous scaffold. Artificial Organs. 2017;**41**(11):E296-E307. DOI: 10.1111/aor.12926

[78] Dzikowski M, Castanié N, Guedon A, Verrier B, Primard C, Sohier J. Antibiotic incorporation in jet-sprayed nanofibrillar biodegradable scaffolds for wound healing. International Journal of Pharmaceutics. 2017;**532**(2):802-812. DOI: 10.1016/j.ijpharm.2017.08.117

[79] Steffens D, Leonardi D, Soster PR, Lersch M, Rosa A, Crestani T, et al. Development of a new nanofiber scaffold for use with stem cells in a third degree burn animal model. Burns. 2014;**40**(8):1650-1660. DOI: 10.1016/j.burns.2014.03.008

[80] Mohiti-Asli M, Saha S, Murphy SV, Gracz H, Pourdeyhimi B, Atala A, et al. Ibuprofen loaded PLA nanofibrous scaffolds increase proliferation of human skin cells in vitro and promote healing of full thickness incision wounds in vivo. Journal of Biomedical Materials Research. Part B, Applied Biomaterials. 2017;**105**(2):327-339. DOI: 10.1002/jbm.b.33520

[81] Blackstone BN, Hahn JM, McFarland KL, DeBruler DM, Supp DM, Powell HM. Inflammatory response and biomechanical properties of coaxial scaffolds for engineered skin in vitro and post-grafting. Acta Biomaterialia. 2018;**80**:247-257. DOI: 10.1016/j.actbio.2018.09.014

[82] Yaru W, Lan X, Jianhua S, Chenxu F. Preparation, characterization and drug release of salicylic acid loaded porous electrospun nanofibers. Recent Patents on Nanotechnology. 2018;**12**(3):208-217. DOI: 10.2174/1872210512666181029154908

[83] Wu KK. Salicylates and their spectrum of activity. Anti-Inflammatory & Anti-Allergy Agents in Medicinal Chemistry. 2007;**6**(4):278-292. DOI: 10.2174/187152307783220031

[84] Chandrasekaran AR, Venugopal J, Sundarrajan S, Ramakrishna S. Fabrication of a nanofibrous scaffold with improved bioactivity for culture of human dermal fibroblasts for skin regeneration. Biomedical Materials. 2011;**6**(1):015001. DOI: 10.1088/1748-6041/6/1/015001

[85] Gandhimathi C, Venugopal JR, Bhaarathy V, Ramakrishna S, Kumar SD. Biocomposite nanofibrous strategies for the controlled release of biomolecules for skin tissue regeneration. International Journal of Nanomedicine. 2014;**9**:4709-4722. DOI: 10.2147/IJN.S65335

[86] Sridhar S, Venugopal JR, Ramakrishna S. Improved regeneration potential of fibroblasts using ascorbic acid-blended nanofibrous scaffolds. Journal of Biomedical Materials Research. Part A. 2015;**103**(11):3431-3440. DOI: 10.1002/jbm.a.35486

[87] Gu J, Liu N, Yang X, Feng Z, Qi F. Adiposed-derived stem cells seeded on PLCL/P123 eletrospun nanofibrous scaffold enhance wound healing. Biomedical Materials. 2014;**9**(3):035012. DOI: 10.1088/1748-6041/9/3/035012

[88] Riedel T, Brynda E, Dyr JE, Houska M. Controlled preparation of thin fibrin films immobilized at solid surfaces. Journal of Biomedical Materials Research. Part A. 2009;**88**(2):437-447. DOI: 10.1002/jbm.a.31755

[89] Kao CW, Tseng YY, Liu KS, Liu YW, Chen JC, He HL, et al. Anesthetics and human epidermal growth factor incorporated into anti-adhesive nanofibers provide sustained pain relief and promote healing of surgical wounds. International Journal of Nanomedicine. 2019;**14**:4007-4016. DOI: 10.2147/IJN.S202402

[90] Mirdailami O, Soleimani M, Dinarvand R, Khoshayand MR, Norouzi M, Hajarizadeh A, et al. Controlled release of rhEGF and rhbFGF from electrospun scaffolds for skin regeneration. Journal of Biomedical Materials Research. Part A. 2015;**103**(10):3374-3385. DOI: 10.1002/jbm.a.35479

[91] Zheng Z, Liu Y, Huang W, Mo Y, Lan Y, Guo R, et al. Neurotensin-loaded PLGA/CNC composite nanofiber membranes accelerate diabetic wound healing. Artificial Cells, Nanomedicine, and Biotechnology. 2018;**46**(suppl 2):493-501. DOI: 10.1080/21691401.2018.1460372

[92] Mo Y, Guo R, Zhang Y, Xue W, Cheng B, Zhang Y. Controlled dual delivery of Angiogenin and curcumin by electrospun nanofibers for skin regeneration. Tissue Engineering. Part A. 2017;**23**(13-14):597-608. DOI: 10.1089/ten.tea.2016.0268

[93] Shtrichman R, Zeevi-Levin N, Zaid R, Barak E, Fishman B, Ziskind A, et al. The generation of hybrid electrospun nanofiber layer with extracellular matrix derived from human pluripotent stem cells, for regenerative medicine applications. Tissue Engineering. Part A. 2014;**20**(19-20):2756-2767. DOI: 10.1089/ten.TEA.2013.0705

[94] Alamein MA, Stephens S, Liu Q, Skabo S, Warnke PH. Mass production of nanofibrous extracellular matrix with controlled 3D morphology for large-scale soft tissue regeneration. Tissue Engineering. Part C, Methods. 2013;**19**(6):458-472. DOI: 10.1089/ten.TEC.2012.0417

[95] Poormasjedi-Meibod MS, Salimi Elizei S, Leung V, Baradar Jalili R, Ko F, Ghahary A. Kynurenine modulates MMP-1 and type-I collagen expression via aryl hydrocarbon receptor activation in dermal fibroblasts. Journal of Cellular Physiology. 2016;**231**(12):2749-2760. DOI: 10.1002/jcp.25383

[96] Li J, Fu R, Li L, Yang G, Ding S, Zhong Z, et al. Co-delivery of dexamethasone and green tea polyphenols using electrospun ultrafine fibers for effective treatment of keloid. Pharmaceutical Research. 2014;**31**(7):1632-1643. DOI: 10.1007/s11095-013-1266-2

[97] Song J, Fan X, Shen Q. Daidzein-loaded nanostructured lipid carriers-PLGA nanofibers for transdermal delivery. International Journal of Pharmaceutics. 2016;**501**(1-2):245-252. DOI: 10.1016/j.ijpharm.2016.02.003

[98] Naves L, Dhand C, Almeida L, Rajamani L, Ramakrishna S, Soares G. Poly(lactic-co-glycolic) acid drug delivery systems through transdermal pathway: An overview. Progress in Biomaterials. 2017;**6**(1-2): 1-11. DOI: 10.1007/s40204-017-0063-0

[99] Selvaraj S, Thangam R, Fathima NN. Electrospinning of casein nanofibers with silver nanoparticles for potential biomedical applications. International Journal of Biological Macromolecules. 2018;**120**(Pt B):1674-1681. DOI: 10.1016/j.ijbiomac.2018.09.177

[100] Lin S, Chen M, Jiang H, Fan L, Sun B, Yu F, et al. Green electrospun grape seed extract-loaded silk fibroin nanofibrous mats with excellent cytocompatibility and antioxidant effect. Colloids Surf B Biointerfaces. 2016;**139**:156-163. DOI: 10.1016/j.colsurfb.2015.12.001

[101] Kurečič M, Rijavec T, Hribernik S, Lapanje A, Kleinschek KS, Maver U. Novel electrospun fibers with incorporated commensal bacteria for potential preventive treatment of the diabetic foot. Nanomedicine (London, England). 2018;**13**(13):1583-1594. DOI: 10.2217/nnm-2018-0014

[102] Zhang H, Xia J, Pang X, Zhao M, Wang B, Yang L, et al. Magnetic nanoparticle-loaded electrospun polymeric nanofibers for tissue engineering. Materials Science & Engineering. C, Materials for Biological Applications. 2017;**73**:537-543. DOI: 10.1016/j.msec.2016.12.116

[103] Loh XJ, Peh P, Liao S, Sng C, Li J. Controlled drug release from biodegradable thermoresponsive physical hydrogel nanofibers. Journal of Controlled Release. 2010;**143**(2):175-182. DOI: 10.1016/j.jconrel.2009.12.030

[104] Pan JF, Liu NH, Shu LY, Sun H. Application of avidin-biotin technology to improve cell adhesion on nanofibrous matrices. Journal of Nanobiotechnology. 2015;**13**:37. DOI: 10.1186/s12951-015-0096-2

[105] Aldana AA, Abraham GA. Current advances in electrospun gelatin-based scaffolds for tissue engineering applications. International Journal of Pharmaceutics. 2017;**523**(2):441-453. DOI: 10.1016/j.ijpharm.2016.09.044

[106] Bannerman AD, Li X, Wan W. A 'degradable' poly(vinyl alcohol) iron oxide nanoparticle hydrogel. Acta Biomaterialia. 2017;**58**:376-385. DOI: 10.1016/j.actbio.2017.05.018

[107] Shamaz BH, Anitha A, Vijayamohan M, Kuttappan S, Nair S, Nair MB. Relevance of fiber integrated gelatin-nanohydroxyapatite composite scaffold for bone tissue regeneration. Nanotechnology. 2015;**26**(40):405101. DOI: 10.1088/0957-4484/26/40/405101

[108] Wu J, Hong Y. Enhancing cell infiltration of electrospun fibrous scaffolds in tissue regeneration. Bioactive Materials. 2016;**1**(1):56-64. DOI: 10.1016/j.bioactmat.2016.07.001

[109] Zhao X, Sun X, Yildirimer L, Lang Q, Lin ZYW, Zheng R, et al. Cell infiltrative hydrogel fibrous scaffolds for accelerated wound healing. Acta Biomaterialia. 2017;**49**:66-77. DOI: 10.1016/j.actbio.2016.11.017

[110] Zarekhalili Z, Bahrami SH, Ranjbar-Mohammadi M, Milan PB. Fabrication and characterization of PVA/Gum tragacanth/PCL hybrid nanofibrous scaffolds for skin substitutes. International Journal of Biological Macromolecules. 2017;**94**(Pt A):679-690. DOI: 10.1016/j.ijbiomac.2016.10.042

[111] Safaee-Ardakani MR, Hatamian-Zarmi A, Sadat SM, Mokhtari-Hosseini ZB, Ebrahimi-Hosseinzadeh B, Rashidiani J, et al. Electrospun Schizophyllan/polyvinyl alcohol blend nanofibrous scaffold as potential wound healing. International Journal of Biological Macromolecules. 2019;**127**:27-38. DOI: 10.1016/j.ijbiomac.2018.12.256

[112] Aghamohamadi N, Sanjani NS, Majidi RF, Nasrollahi SA. Preparation and characterization of Aloe vera acetate and electrospinning fibers as promising antibacterial properties materials. Materials Science & Engineering. C, Materials for Biological Applications. 2019;**94**:445-452. DOI: 10.1016/j.msec.2018.09.058

[113] Ajalloueian F, Tavanai H, Hilborn J, Donzel-Gargand O, Leifer K, Wickham A, et al. Emulsion electrospinning as an approach to fabricate PLGA/chitosan nanofibers for biomedical applications. BioMed Research International. 2014;**2014**:475280. DOI: 10.1155/2014/475280

[114] Contardi M, Heredia-Guerrero JA, Perotto G, Valentini P, Pompa PP, Spanò R, et al. Transparent ciprofloxacin-povidone antibiotic films and nanofiber mats as potential skin and wound care dressings. European Journal of Pharmaceutical Sciences. 2017;**104**:133-144. DOI: 10.1016/j.ejps.2017.03.044

[115] Tamm I, Heinämäki J, Laidmäe I, Rammo L, Paaver U, Ingebrigtsen SG, et al. Development of suberin fatty acids and chloramphenicol-loaded antimicrobial electrospun Nanofibrous Mats intended for wound therapy. Journal of Pharmaceutical Sciences. 2016;**105**(3):1239-1247. DOI: 10.1016/j.xphs.2015.12.025

[116] Shankhwar N, Kumar M, Mandal BB, Robi PS, Srinivasan A. Electrospun polyvinyl alcohol-polyvinyl pyrrolidone nanofibrous membranes for interactive wound dressing application. Journal of Biomaterials Science. Polymer Edition. 2016;**27**(3):247-262. DOI: 10.1080/09205063.2015.1120474

[117] Shi Y, Zhang J, Xu S, Dong A. Electrospinning of artemisinin-loaded core-shell fibers for inhibiting drug re-crystallization. Journal of Biomaterials Science. Polymer Edition. 2013;**24**(5):551-564. DOI: 10.1080/09205063.2012.698895

[118] Vitacolonna M, Belharazem D, Hohenberger P, Roessner ED. Effect of dynamic seeding methods on the distribution of fibroblasts within human acellular dermis. Cell and Tissue Banking. 2015;**16**(4):605-614. DOI: 10.1007/s10561-015-9508-7

[119] Lei XH, Ning LN, Cao YJ, Liu S, Zhang SB, Qiu ZF, et al. NASA-approved rotary bioreactor enhances proliferation of human epidermal stem cells and supports formation of 3D epidermis-like structure. PLoS One. 2011;**6**(11):e26603. DOI: 10.1371/journal.pone.0026603

[120] Bye FJ, Wang L, Bullock AJ, Blackwood KA, Ryan AJ, MacNeil S. Postproduction processing of electrospun fibres for tissue engineering. Journal of Visualized Experiments. 2012;**66**:4172. DOI: 10.3791/4172

[121] Renò F, Traina V, Cannas M. Mechanical stretching modulates growth direction and MMP-9 release in human keratinocyte monolayer. Cell Adhesion & Migration. 2009;**3**(3):239-242. DOI: 10.4161/cam.3.3.8632

[122] Hossain MM, Crish JF, Eckert RL, Lin JJ, Jin JP. h2-Calponin is regulated by mechanical tension and modifies the function of actin cytoskeleton. The Journal of Biological Chemistry. 2005;**280**(51):42442-42453. DOI: 10.1074/jbc.M509952200

[123] Tsutsumi M, Inoue K, Denda S, Ikeyama K, Goto M, Denda M. Mechanical-stimulation-evoked calcium waves in proliferating and differentiated human keratinocytes. Cell and Tissue Research. 2009;**338**(1):99-106. DOI: 10.1007/s00441-009-0848-0

[124] Skogberg A, Mäki AJ, Mettänen M, Lahtinen P, Kallio P. Cellulose nanofiber alignment using evaporation-induced droplet-casting, and cell alignment on aligned nanocellulose surfaces. Biomacromolecules. 2017;**18**(12): 3936-3953. DOI: 10.1021/acs.biomac.7b00963

[125] Kumamoto J, Goto M, Nagayama M, Denda M. Real-time imaging of human epidermal calcium dynamics in response to point laser stimulation. Journal of Dermatological Science. 2017;**86**(1):13-20. DOI: 10.1016/j.jdermsci.2017.01.002

[126] Arai KY, Nakamura Y, Hachiya Y, Tsuchiya H, Akimoto R, Hosoki K, et al. Pulsed electric current induces the differentiation of human keratinocytes. Molecular and Cellular Biochemistry. 2013;**379**(1-2):235-241. DOI: 10.1007/s11010-013-1645-3

[127] Jang HK, Oh JY, Jeong GJ, Lee TJ, Im GB, Lee JR, et al. A disposable photovoltaic patch controlling cellular microenvironment for wound healing. International Journal of Molecular Sciences. 2018;**19**(10):3025. DOI: 10.3390/ijms19103025

[128] Jin G, Li J, Li K. Photosensitive semiconducting polymer-incorporated nanofibers for promoting the regeneration of skin wound. Materials Science & Engineering. C, Materials for Biological Applications. 2017;**70**(Pt 2):1176-1181. DOI: 10.1016/j.msec.2016.04.107

[129] Ito A, Hayashida M, Honda H, Hata K, Kagami H, Ueda M, et al. Construction and harvest of multilayered keratinocyte sheets using magnetite nanoparticles and magnetic force. Tissue Engineering. 2004;**10**(5-6):873-880. DOI: 10.1089/1076327041348446

[130] Koria P, Brazeau D, Kirkwood K, Hayden P, Klausner M, Andreadis ST. Gene expression profile of tissue engineered skin subjected to acute barrier disruption. The Journal of Investigative Dermatology. 2003;**121**(2):368-382. DOI: 10.1046/j.1523-1747.2003.12364.x

[131] Fatimah SS, Chua K, Tan GC, Azmi TI, Tan AE, Abdul RH. Organotypic culture of human amnion cells in air-liquid interface as a potential substitute for skin regeneration. Cytotherapy. 2013;**15**(8):1030-1041. DOI: 10.1016/j.jcyt.2013.05.003

[132] Petry L, Kippenberger S, Meissner M, Kleemann J, Kaufmann R, Rieger UM, et al. Directing adipose-derived stem cells into keratinocyte-like cells: Impact of medium composition and culture condition. Journal of the European Academy of Dermatology and Venereology. 2018;**32**(11):2010-2019. DOI: 10.1111/jdv.15010

[133] Lu H, Rollman O. Fluorescence imaging of reepithelialization from skin explant cultures on acellular dermis. Wound Repair and Regeneration. 2004;**12**(5):575-586. DOI: 10.1111/j.1067-1927.2004.012510.x

[134] Wang TW, Huang YC, Sun JS, Lin FH. Organotypic keratinocyte-fibroblast cocultures on a bilayer gelatin scaffold as a model of skin equivalent. Biomedical Sciences Instrumentation. 2003;**39**:523-528

[135] John S, Kesting MR, Paulitschke P, Stöckelhuber M, von Bomhard A. Development of a tissue-engineered skin substitute on a base of human amniotic membrane. Journal of Tissue Engineering. 2019;**10**:1-14. DOI: 10.1177/2041731418825378

[136] Keck M, Gugerell A, Kober J. Engineering a multilayered skin substitute with keratinocytes, fibroblasts, adipose-derived stem cells, and adipocytes. Methods in Molecular Biology. 1993;**2019**:149-157. DOI: 10.1007/978-1-4939-9473-1_12

[137] Liu Q, Zhang RZ, Xu B. Construction of a three-dimensional in-vitro skin model on polycaprolactone fibers. Giornale Italiano Di Dermatologia E Venereologia. 2018;**153**(5):636-643. DOI: 10.23736/S0392-0488.17.05472-4

[138] Planz V, Wang J, Windbergs M. Establishment of a cell-based wound healing assay for bio-relevant testing of wound therapeutics. Journal of Pharmacological and Toxicological Methods. 2018;**89**:19-25. DOI: 10.1016/j.vascn.2017.10.003

[139] Pouliot R, Larouche D, Auger FA, Juhasz J, Xu W, Li H, et al. Reconstructed human skin produced in vitro and grafted on athymic mice. Transplantation. 2002;**73**(11):1751-1757. DOI: 10.1097/00007890-200206150-00010

[140] Kalyanaraman B, Supp DM, Boyce ST. Medium flow rate regulates viability and barrier function of engineered skin substitutes in perfusion culture. Tissue Engineering. Part A. 2008;**14**(5):583-593. DOI: 10.1089/tea.2007.0237

Chapter 4

# Applications of Nanotechnology in Agriculture

*Alaa Y. Ghidan and Tawfiq M. Al Antary*


## Abstract

Nanotechnology has gained intense attention in the recent years due to its wide applications in several areas like medicine, medical drugs, catalysis, energy and materials. Those nanoparticles with small size to large surface area (1–100 nm) have several potential functions. These days, sustainable agriculture is needed. The development of nanochemicals has appeared as promising agents for the plant growth, fertilizers and pesticides. In recent years, the use of nanomaterials has been considered as an alternative solution to control plant pests including insects, fungi and weeds. Several nanomaterials are used as antimicrobial agents in food packing in which several nanoparticles such as silver nanomaterials are in great interest. Many nanoparticles (Ag, Fe, Cu, Si, Al, Zn, ZnO, $TiO_2$, $CeO_2$, $Al_2O_3$ and carbon nanotubes) have been reported to have some adverse effects on plant growth apart from the antimicrobial properties. In food industries, nanoparticles are leading in forming the food with high quality and good nutritive value.


**Keywords:** agriculture, food industries, applications, nanoparticles, pesticides, fertilizers, antimicrobial

## 1. Introduction

Nanotechnology has gained intense attention in recent years due to its wide applications in several areas like medicine, medical drugs, catalysis, energy and materials. Those nanoparticles with small size to large surface area (1–100 nm) have potential medical, industrial and agricultural applications. Scientists have carried out significant efforts toward the synthesis of nanoparticles by different means, including physical, chemical and biological methods [1]. These methods have many disadvantages due to the difficulty of scale-up of the process, separation and purification of nanoparticles from the micro-emulsion (oil, surfactant, co-surfactant and aqueous phase) and consuming large amount of surfactants [2]. Green methods for synthesizing nanoparticles with plant extracts are advantageous as it is simple, convenient, eco-friendly and require less reaction time. Nanomaterials prepared by eco-friendly and green methods could increase agriculture potential for improving the fertilization process, plant growth regulators and pesticides [3]. In addition, they minimize the amount of harmful chemicals that pollutes the environment. Hence, this technology helps in reducing the environmental pollutants [4], and nanotechnology has recently gained attention due to its wide applications in different fields such as in medicine, environment and agriculture [5]. Particularly, the large surface area offered by the tiny nanoparticles, which have high surface area, makes them attractive to address challenges not met by physical, chemical pesticides and biological control methods.

Nanotechnology in agriculture has gained good momentum in the last decade with an abundance of public funding, but the stage of development is good, even though many methods became under the umbrella of agriculture. This might be attributed to a unique nature of farm production, which functions as an open system whereby energy and matter are exchanged freely. The scale of demand of input materials is always being large in contrast with industrial nanoproducts with the absence of control over the input of the nanomaterials in contrast with industrial nanoproducts [6]. Nanotechnology provides new agrochemical agents and new delivery mechanisms to improve crop productivity, and it promises to reduce pesticide applications. Nanotechnology can increase agricultural production, and its applications include: (1) nanoformulations of agrochemicals for applying pesticides and fertilizers for crop improvement; (2) the application of nanosensors in crop protection for the identification of diseases and residues of agrochemicals; (3) nanodevices for the genetic engineering of plants; (4) plant disease diagnostics; (5) animal health, animal breeding, poultry production; and (6) postharvest management. Precision farming techniques might be used to further improve the crop yields but not damage soil and water. In addition, it can reduce nitrogen loss due to leaching and emissions, and soil microorganisms. Nanotechnology applications include nanoparticle-mediated gene or DNA transfer in plants for the development of insect-resistant varieties, food processing and storage and increased product shelf life. Nanotechnology may increase the development of biomass-to-fuel production. Experts feel that the potential benefits of nanotechnology for agriculture, food, fisheries and aquaculture need to be balanced against concerns for the soil, water and environment and the occupational health of workers [7]. Nanotechnology uses are currently being researched, tested and in some cases already applied in food technology [8]. Nanomaterials are considered with specific chemical, physical and mechanical properties. In recent years, agricultural waste products have attracted attention as source of renewable raw materials to be processed in substitution of several different applications as well as a raw material for nonmaterial production. Insecticide resistance is one of the best examples of evolution occurring on an ecological time scale. The study of insecticide resistance is needed, both because it leads to understanding mechanisms operating in real time and because of its economic importance. It has become in insects an increasing problem for agriculture and public health. Agricultural practices could include wide range of selective regimes [1]. Nanotechnology applications are being tested in food technology and agriculture. The applications of nanomaterials in agriculture aim to reduce spraying of plant protection products and to increase plant yields. Nanotechnology means like nanocapsules, and nanoparticles are examples of uses for the detection and treatment of diseases. Nanotechnology derived devices are also explored in the field of plant breeding and genetic transformation. The potential of nanotechnology in agriculture is large, but a few issues are still to be addressed as the risk assessment. In this respect, some nanoparticle attractants are derived from biopolymers such as proteins and carbohydrates with low effect on human health and the environment. Nanotechnology has many uses in all stages of production, processing, storing, packaging and transport of agricultural products. Nanotechnology will revolutionize agriculture and food industry such as in case of farming techniques, enhancing the ability of plants to absorb nutrients, disease detection and control pests.

## 2. Nanotechnology in pesticides and fertilizers

These days, sustainable agriculture is needed. It may be understood to present a good approach of ecosystem for long run. Practices that can cause long-term damage to *soil* include excessive *tilling* of the soil which leads to *erosion* and *irrigation* without needed drainage. This will lead to *salinization*. This is to satisfy human being food, animal feed and fiber needs.

*Long-term experiments* are required to show the effect of different practices on soil properties which are essential to sustainability and to provide important data on this objective. In the United States, a federal agency, the development of nano-chemicals has appeared as promising agents for the plant growth and pest control. The fertilizers are required in plants growth. Nanomaterials act as fertilizers might have the properties such as crop improvement and with less eco-toxicity. Plants can give an important way for their bioaccumulation into the food chain. The recent developments in agriculture cover the applications of NPs for more effective and safe use of chemicals for plants. The effects of different NPs on plant growth and phytotoxicity were reported by several workers including magnetite ($Fe_3O_4$) nanoparticles and plant growth [9], alumina, zinc, and zinc oxide on seed germination and root growth of five higher plant species; radish, rape, lettuce, corn, and cucumber, silver nanoparticles and seedling growth in wheat [10], sulfur nanoparticles on tomato [11], zinc oxide in mungbean, nanoparticles of AlO, CuO, FeO, MnO, NiO, and ZnO [12]. Silver nanoparticles can stimulate wheat growth and yield. Soil applied 25 ppm SNPs had highly favorable growth promoting effects on wheat growth and yield.

Zinc has been considered as an essential micronutrient for metabolic activities in plants although it is required in trace amounts in plants. It was found that zinc has an important role in management of reactive oxygen species and protection of plant cells against oxidative stresses. Zinc has important functions in the synthesis of auxin or indoleacetic acid (IAA) from tryptophan as well as in biochemical reactions required for formation of chlorophyll and carbohydrates. The crop yield and quality of produce can be affected by deficiency of Zn. The development of insecticide resistance in pest insects has been an increasing problem for agriculture and public health.

Magnesium oxide (MgO) is important inorganic materials with many uses such as adsorbents, fire retardants, advanced ceramics, toxic waste remediation, and photo electronic materials. Therefore, various techniques and routes for synthesis of MgONPs have been reported [1]. MgOH was synthesized by green methods using nontoxic *neem* leaves extract [13], *Citrus limon* leaves extract, *acacia* gum [14].

### 2.1 Control of plant pests

Fusarium wilt is a destructive disease of tomato and lettuce in several countries due to its severe production loss, prolonged survival of fungus in soil and generation of resistant races. The disease can be reduced to some extent with the use of resistant cultivars and chemicals. However, the occurrence and development of new pathogenic races is a continuing problem, and the use of chemicals is expensive and not always effective. In recent years, the use of nanomaterials has been considered as an alternative solution to control plant pathogens. Ghidan et al. [15] has synthesized nanoparticles of magnesium oxide (MgO) and tested the effect of different concentrations on the green peach aphid (GPA) under the greenhouse conditions [16–19]. The synthesis of nanomaterials of copper oxide (CuO), zinc oxide (ZnO), magnesium hydroxide (MgOH) and magnesium oxide (MgO) has been carried out successfully by using aqueous extracts of *Punica granatum* peels, *Olea europaea* leaves and *Chamaemelum nobile* flowers [1]. The screening of synthesized bio-nanoparticles revealed that these nanoparticles were effective in increasing the mortality percent of green peach aphid. After the glasshouse experiments, the metal oxide nanoparticle accumulations were analyzed in the fruits and leaves of green sweet pepper. The results showed that there was no any metal accumulation in any of the plant fruits. Foliar spray by synthesized of MgOH nanoparticles for green pepper leaves revealed that the foliar spraying leaves with 100–800 ppm metal nanoparticles are very beneficial to plant growth and produced healthy plants with greener leaves and high fruit quality compared to the control. Researchers made significant efforts toward the synthesis of nanoparticles by various means, including physical, chemical and biological methods [1]. Green methods for synthesizing nanoparticles with plant extracts are advantageous as it is simple, convenient, environment friendly and require less reaction time. Nanomaterials prepared by eco-friendly and green methods may increase agriculture potential for improving the fertilization process, plant growth and pesticides. In addition, this technology minimizes the amount of harmful chemicals that pollutes the environment [4]. The green peach aphid is considered as a key pest on peach and globally important pest of a broad range of arable and horticultural crops, including Jordan. The pest is categorized as of the most important agricultural pest in the world. This devastated pest combats organophosphorus and carbamate insecticides by overproducing insecticide-degrading carboxyl esterases. Moreover, control of such a pest is becoming increasingly difficult, because the overproduction of resistance for aphid individuals when using chemical insecticides such as carbamates, organophosphates and pyrethroids [20].

Nanomaterials such as copper oxide (CuONPs), zinc oxide (ZnONPs), magnesium hydroxide (MgOHNPs) and magnesium oxide (MgONPs) were synthesized by different physical and chemical methods [21]. With the growing needs to minimize the use of environmental-risk substances, such as insecticides, the biosynthesis of nanoparticles as an emerging highlight of the intersection of nanotechnology and biotechnology has received increasing attention. The rate of reduction of metal ions using plants has been found to be much faster as compared to microorganisms and stable formation of nanoparticles has been reported.

## 2.2 Nanoinsicticidal potential

Copper oxide nanoparticles (CuONPs) are synthesized through different methods [22] such as precipitation [23] and chemical reduction [24]. Many plant aqueous extracts have been reported such as *Citrus limon* juice [25] and carob leaves [26]. Applications led many researchers to develop different ways to synthesis ZnONPs such as chemical route [27], precipitation method [28], hydrolyzed in polar organic solvents [29] and microwave synthesis [30]. Different plant extracts have been reported in the open literature for green synthesis of ZnONPs such as *Olea europaea*, *Solanum nigrum* leaves [31] and *Azadirachta indica* [20]. Different methods for synthesis MgOHNPs and MgONPs have been reported such as hydrothermal route, water-in-oil microemulsion and microwave reaction [32]. MgOH was synthesized by green methods using nontoxic and eco-friendly such as *Neem* leaves extract, *Citrus limon* leaves extract, *acacia* gum, *Brassica oleracea* and *Punica granatum* peels [3, 22]. In agriculture sector, there are several uses available like nanotech based pesticides and fertilizers with effective impact on plant growth and molecular farming with the help of nanovectors which is hoping to take the place of viral vectors [33].

## 2.3 Antimicrobial activity

Several nanomaterials are used as antimicrobial agents in food packing in which silver nanoparticles are in great interest. This is because of its extended use. Some other nanoparticles currently used are titanium dioxide ($TiO_2$), zinc oxide (ZnO), silicon oxide ($SiO_2$), magnesium oxide (MgO), gold and silver. All of them have specific characteristics and functions, for example, zinc nanocrystal shows antimicrobial and antifungal activity [34]. Silver was a disinfectant and sterilizing agent used by NASA and Russian Space station for water [35], silver zeolite and silver. Gold has high temperature stability and low volatility and good antifungal and antimicrobial effects against 150 different bacteria [36]. FDA in 2009 approves the direct use of silver as disinfectant in commercial water, since with effective result against microorganisms. The antimicrobial effect of these are *E. coli*, *L. monocytogenes* and *Staphylococcus aureus* [37], and nanosilver particles coated with cellulose acetate phthalate also provided similar results [37]. Some nanoparticles have shown their antifungal activity. These fungi include *Candida albicans*, *Aspergillus niger* [38] and yeast [39]. AgNPs are also found to be effective against methicillin resistant *Staphylococcus aureus* [39]. Other nanoparticles besides silver are also found to have antimicrobial characteristics like titanium oxide ($TiO_2$). Its antimicrobial activity in UV light was obvious. Zinc oxide is reported to have antimicrobial activity in packaging material [39].

Zinc oxide nanoparticles synthesized using *Punica granatum* peel aqueous extract has shown effectiveness as antibacterial agents against standard strains of Gram-positive *Staphylococcus aureus* and Gram-negative *Escherichia coli* [1, 40].

### 2.4 Nanotechnology application as nanofungicides

The use of nanosilver has been studied recently against phytopathogen *Colletotrichum gloeosporioides* [41]. Other nanoparticles (Fe, Cu, Si, Al, Zn, ZnO, $TiO_2$, $CeO_2$, $Al_2O_3$ and carbon nanotubes) have been reported to have some adverse effects on plant growth apart from the antimicrobial properties [42]. Sometimes, nanoparticles also have an effect on the growth of useful soil bacteria, such as *Pseudomonas putida KT2440* [43]. Various research groups focused their interest on the usage of eco-friendly pesticides. Similar to chemical pesticides, nanoparticle-based pesticides and herbicides are being explored for the application of the antimicrobial agents to protect crops from various diseases. Extensive studies on nanoparticle-based systems may eliminate the intensive use of pesticides in the agricultural sector [44]. The antifungal properties of nanoparticles can help to formulate nanoparticle-based pesticides [41]. Among the different inorganic nanoparticle-based antimicrobial agents, silver has been extensively studied by many researchers because of its several advantages over other nanoparticles such as copper, zinc, gold, ZnO, $Al_2O_3$ and $TiO_2$.

#### *2.4.1 Effect of nickel nanoparticles on fungal mycelial growth*

Both the concentrations of nickel nanoparticles (50 and 100 ppm) inhibited the fungal mycelial growth on solid media, and the inhibitions were significant ($p \leq 0.05$) over control (**Figure 1**). Nickel nanoparticles at 100 ppm concentration inhibited the mycelial growth of *F. oxysporum* f. sp. *lactucae* and *F. oxysporum* f. sp. *lycopersici* by 60.23 and 59.77%, respectively, over control.

The inhibitory effects of nickel nanoparticles were also assessed in liquid medium, and the results were similar with solid media. In the liquid media, the fresh mycelial weight of the tested fungal pathogens decreased significantly and more than 50% reduction was recorded with the use of nickel nanoparticles at the concentration of 100 ppm. The results revealed that mycelial growth of tested pathogens was inhibited in a concentration dependent manner. These results

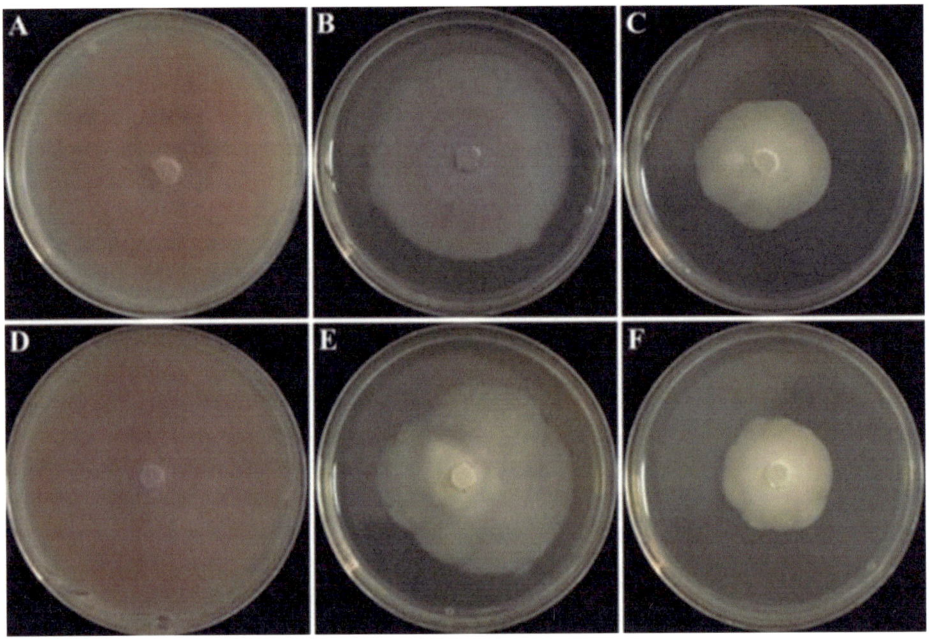


**Figure 1.**
*Petri dishes showing inhibition of Fusarium wilt causing pathogens: first row, Fusarium oxysporum f. sp. lycopersici [(A) control; (B) 50 ppm nickel nanoparticles; (C) 100 ppm nickel nanoparticles]; second row, Fusarium oxysporum f. sp. lactucae [(D) control; (E) 50 ppm nickel nanoparticles; (F) 100 ppm nickel nanoparticles] [45].*

suggest that using Ni nanoparticle solution can significantly increase the surface areas acting on the mycelia of *Fusarium* and mycelial growth.

Nickel nanoparticles at the concentration of 100 ppm decreased the number of spore development by 81.40 and 74.60% in *F. oxysporum* f. sp. *lactucae* and *F. oxysporum* f. sp. *lycopersici*, respectively. The conidial germination was negatively affected by nickel nanoparticles (**Figure 2**).

The inhibitory effect of Ni nanoparticles on spore germination could be due to their fungicidal effect. These results agreed with the results obtained from other workers on antifungal effects of different metal nanoparticles against some pathogenic fungi such as silver nanoparticles and zinc nanoparticles against copper nanoparticles [46]. Inhibitory effect of Ni nanoparticles could be due to producing of extracellular enzymes from fungi as survival agents caused by stress of toxic materials [47] or could be due to large surface areas (**Figure 3**) and small sizes to penetrate into the cell membrane of pathogen and work in the cytosols [48].

## 2.5 Nanotechnology for controlling plant virus

Plant virus particularly spherical virus is considered to be the naturally occurring nanomaterials. The smallest plant viruses known till date are satellite tobacco necrosis virus measuring only 18 nm in diameter [49]. Plant viruses are made up of single or double stranded RNA/DNA as genome which is encapsulated by a protein coat. Their ability to infect, deliver nucleic acid genome to a specific site in host cell, replicate, package nucleic acid and come out of host cell precisely in an orderly manner have necessitated them to be used in nanotechnology. A complete review on use of plant viruses as bio templates for nanomaterials and their uses has been done recently by Young et al. [50].

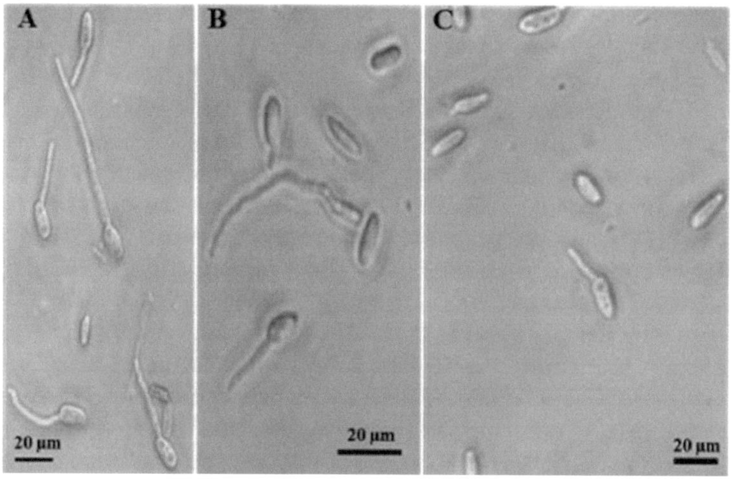


**Figure 2.**
*Germination of Fusarium oxysporum f. sp. lycopersici conidia: (A) control; (B) 50 ppm nickel nanoparticles; and (C) 100 ppm nickel nanoparticles [45].*

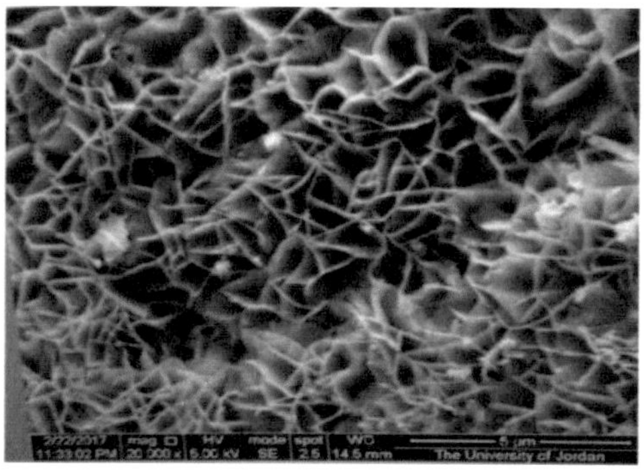

**Figure 3.**
*SEM, nanoflex of large surface areas of MgONPs [1].*

## 3. Nanotechnology in food packaging

Food industries are leading in forming the food with good nutritive value. For example, high impermeable packaging nanomaterials are used for protection of food from UV radiations and providing more strength to maintain the food protected from environment, increasing their shelf lives. Nanosensors are used for the detection of chemicals, gases and pathogens in food. In modern terminology, a word is given to such type of packaging as smart packaging. Some studies suggested that people are not accepting the direct involvement of nanoparticles in food due to some risk factors. Therefore, it is needed to provide some safety measurements to reduce the risk and human safety.

## 4. Conclusions

The occurrence and development of new pathogenic races is a continuing problem, and the use of chemicals to control pests is expensive and not always effective. In recent years, the use of nanomaterials has been considered as an alternative solution to control plant pathogens. Agricultural practices usually include the systematic application of a wide array of active compounds at variable dosages and frequencies, which represent a wide range of selective regimes.

Metal oxide nanoparticles have controlled the green peach aphid. Magnesium hydroxide, bionanoparticles synthesized were the best control to *Myzus persicae* compared to other synthetic nanoparticles. Zinc oxide nanoparticles synthesized using aqueous *Punica granatum* peel extract was tested for its potential antimicrobial activity against some selected microbes. Also this research work determined the effect of synthesized ZnONPs on green peach aphid and antibacterial efficacy against standard strains of Gram-positive *Staphylococcus aureus* and Gram-negative *Escherichia coli*. Other nanoparticles (Fe, Cu, Si, Al, Zn, ZnO, $TiO_2$, $CeO_2$, $Al_2O_3$ and carbon nanotubes) have also been reported to have some adverse effects on plant growth. Sometimes nanoparticles also have an adverse effect on the growth of useful soil bacteria, such as *Pseudomonas putida* KT2440. Both the concentrations of nickel nanoparticles (50 and 100 ppm) inhibited the fungal mycelial growth on solid media, and the inhibitions were significant over control. Nickel nanoparticles at 100 ppm concentration inhibited the mycelial growth of F. *oxysporum* f. sp. *lactucae* and F. *oxysporum* f. sp. *Lycopersici*.

Green methods for synthesizing nanoparticles with plant extracts are advantageous as it is simple, convenient, environment friendly and require less reaction time. Nanomaterials prepared by eco-friendly and green methods may increase agriculture potential for improving the fertilization process, plant growth regulators, pesticides delivery of active component to the desired target sites, treatment of wastewater and also enhancing the absorption of nutrients in plant. In addition, they minimize the amount of harmful chemicals that pollutes the environment. Hence, this technology helps in reducing the environmental pollutants. Nanotechnology has recently gained attention due to wide applications in different fields such as in agriculture medicine and environment. The large surface area offered by the tiny nanoparticles, which have high surface area, makes them attractive to address challenges not met by different control methods.

Nanotechnology applications are currently being researched, tested and in some cases already applied across the entire spectrum of food technology, from agriculture to food processing, packaging and food supplements. They are with unique chemical, physical, and mechanical properties. In recent years, agricultural waste products have attracted attention as source of renewable raw materials. Insecticide resistance is one of the best examples of evolution occurring on an ecological time scale. The study of insecticide resistance is important, because it leads to a better understanding of evolutionary mechanisms operating in real time. The development of insecticide resistance in pest insects has been an increasing problem for agriculture and public health. Agricultural practices usually include the systematic application of a wide array of active compounds at variable dosages and frequencies.

## Acknowledgements

The authors are appreciating the support of the University of Jordan, Agriculture School, Amman, Jordan; Aqaba University of Technology, Pharmacy School, Aqaba, Jordan; and King Abdul Aziz University, Center of Nanotechnology, Jedda, Saudi Arabia Kingdom. In addition, they appreciate the contribution of Ms. Nada Al-Antary for her typing and computer work.

## Conflict of interest

The authors declare no conflict of interest.

**Author details**

Alaa Y. Ghidan[1] and Tawfiq M. Al Antary[2]*

1 Pharmacy School, Aqaba University of Technology, Aqaba, Jordan

2 Faculty of Agriculture, Department of Plant Protection, University of Jordan, Amman, Jordan

*Address all correspondence to: tawfiqalantary@yahoo.com; t.antary@ju.edu.jo

## References


[1] Ghidan AY, Al-Antary TM, Salem NM, Awwad AM. Facile green synthetic route to the zinc oxide (ZnONPs) nanoparticles: Effect on green peach aphid and antibacterial activity. Journal of Agricultural Science. 2017a;**9**(2):131-138

[2] Pilarska A, Wysokowski M, Markiewicz E, Jesionowski T. Synthesis of magnesium hydroxide and its calcinates by a precipitation method with the use of magnesium sulfate and poly (ethylene glycols). Powder Technology. 2013;**235**:148-157

[3] Ghidan AY, Al Antary TM, Awwad AM. Aphidicidal potential of green synthesized magnesium hydroxide nanoparticles using Olea europaea leaves extract. ARPN Journal of Agricultural and Biological Science. 2017b;**12**(10):293-301

[4] Huang S, Wang L, Liu L, Hou Y, Li L. Nanotechnology in agriculture, livestock, and aquaculture in China. A review. Agronomy for Sustainable Development. 2015;**35**:369-400

[5] Kah M, Hofmann T. Nanopesticide research: Current trends and future priorities. Environment International. 2014;**63**:224-235

[6] Mukhopadhyay SS. Nanotechnology in agriculture: Prospects and constraints. Nanotechnology, Science and Applications. 2014;7:63-71

[7] Sekhon BS. Nanotechnology in agri-food production: An overview. Nanotechnology, Science and Applications. 2014;7:31-53

[8] Nanowerk Spotlight. Nanotechnology in agriculture. 2014. Available from: https://www.nanowerk.com/spotlight/spotid=37064.php

[9] Shankramma K, Yallappa S, Shivanna MB, Manjanna J. $Fe_2O_3$ magnetic nanoparticles to enhance *S. lycopersicum* (tomato) plant growth and their biomineralization. Applied Nanoscience. 2016;**6**:983-990. DOI: 10.1007/s13204-015-0510-y

[10] Lin D, Xing B. Phytotoxicity of nanoparticles: Inhibition of seed germination and root growth. Environmental Pollution. 2007;**150**(2):243-250. DOI: 10.1016/j.envpol.2007.01.016

[11] Salem NM, Albanna LS, Abdeen A, Ibrahim OQ, Awwad AI. Sulfur nanoparticles improves root and shoot growth of tomato. Journal of Agricultural Science. 2016;**8**(4):179-185. DOI: 10.5539/jas.v8n4p179

[12] Ghidan AY, Al-Antary TM, Awwad AM, Ghidan OY, Al Araj SE, Ateyyat MA. Comparison of different green synthesized nanomaterials on green peach aphid as aphicidal potential. Fresenius Environmental Bulletin. 2018a, 2018a;**27**(10):7009-7016

[13] Moorthy SK, Ashok CH, Rao KV, Viswanathana C. Synthesis and characterization of MgO nanoparticles by neem leaves through green method. Materials Today: Proceedings. 2015;**2**:4360-4368

[14] Awwad A, Ahmad A. Biosynthesis, characterization, and optical properties of magnesium hydroxide and oxide nanoflakes using *Citrus limon* leaf extract. Arab Journal of Physical Chemistry. 2014;**1**:65-70

[15] Ghidan AY, Al-Antary TM, Awwad AM, Ayad JY. Physiological effect of some nanomaterials on pepper (*Capsicum annuum* L.) plants. Fresenius Environmental Bulletin. 2018b;**27**(11):7872-7878

[16] Al-Antary T, Al-Momany A. Pests of Garden and Home. 1st ed. Cairo: Dar Al-Arabiah for Publications and Distributions; 2013. p. 360

[17] Al- Momany A, Al-Antary T. Pests of Garden and Home. 2nd ed. Amman: Publications of University of Jordan; 2008. p. 518

[18] Al-Antary T, Khader B. Toxicity of four insecticides on longevity and fecundity of three populations of the green peach aphid *Myzus persicae* (Aphididae: Homoptera) for three generations. Jordan Journal of Agricultural Sciences. 2013;**9**(1):52-62

[19] Al-Antary T, Al-Dabbas M, Shaderma A. Effect of UV-radiation on methomyl, oxamyl, and carbosulfan residues in lettuce juice. Fresenius Environmental Bulletin. 2014;**23**(3a):924-928

[20] Madanat H, Al Antary T, Zarqa M. Toxicity of six ethanol plant extracts against the green peach aphid *Myzus persicae* Sulzer (Homopetra: Aphididae). Fresenius Environmental Bulletin. 2016;**25**:706-718

[21] Jiang T, Wang Y, Meng D, Yu M. Facile synthesis and photocatalytic performance of self-assembly CuO microspheres. Superlattices and Microstructures. 2015;**85**:1-6

[22] Ghidan AY, Al-Antary TM, Awwad AM. Green synthesis of copper oxide nanoparticles using *Punica granatum* peels extract: Effect on green peach Aphid. Environmental Nanotechnology, Monitoring and Management. 2016;**6**:95-98

[23] Sahooli M, Sabbaghi S, Saboori R. Synthesis and characterization of mono sized CuO nanoparticles. Materials Letters. 2012;**81**:169-172

[24] Karthik AD, Geetha K. Synthesis of copper precursor, copper and its oxide nanoparticles by green chemical reduction method and its antimicrobial activity. Journal of Applied Pharmaceutical Science. 2013;**3**:16-21

[25] Mohan S, Singh Y, Verna DK, Hassan SH. Synthesis of CuO nanoparticles through green route using *Citrus Limon* juice and its application as nanosorbent for Cr (VI) remediation: Process optimization with RSM and ANN-GA based model. Process Safety and Environmental Protection. 2015;**96**:156-166

[26] Awwad A, Ibrahim Q. Optical and X-ray diffraction characterization of biosynthesis copper oxide nanoparticles using carob leaf extract. Arab Journal of Physical Chemistry. 2015;**2**:20-24

[27] Singh SC, Gopal R. Synthesis of colloidal zinc oxide nanoparticles by pulsed laser ablation in aqueous media. Physica E: Low-dimensional Systems and Nanostructures. 2008;**40**:724-730

[28] An Y, Lian X, Zou Y, Deng D. Synthesis of porous ZnO structure for gas sensor and photocatalytic applications. Colloids and Surfaces A: Physicochemical and Engineering Aspects. 2014;**447**:81-87

[29] Ehlert S, Lunkenbein T, Breu J, Forster S. Facile large-scale synthetic route to monodisperse ZnO nanocrystals. Colloids and Surfaces A: Physicochemical and Engineering Aspects. 2014;**444**:76-80

[30] Sutradhar P, Debbarma M, Saha M. Microwave synthesis of zinc oxide nanoparticles using coffee powder extract and its application for solar cell. Synthesis and Reactivity in Inorganic, Metal-Organic, and Nano-Metal Chemistry. 2016;**46**:1622-1627

[31] Ramesh M, Anbuvannan M, Viruthagiri G. Green synthesis of ZnO nanoparticles using *Solanum nigrum* leaf

extract and their antibacterial activity. Spectrochimica Acta, Part A: Molecular and Biomolecular Spectroscopy. 2015;**136**(Part B):864-870

[32] Yousefi M. Fast method for synthesis magnesium hydroxide nanoparticles, thermal stable and flame retardant poly vinyl alcohol nanocomposite. Journal of Nanotechnology Studies. 2014;**4**:383-388

[33] Chinnamuthu CR, Murugesa BP. Nanotechnology and agro ecosystem. The Madras Agricultural Journal. 2009;**96**(1-6):17-31

[34] Duncan TV. Applications of nanotechnology in food packaging and food safety: Barrier materials, antimicrobials and sensors. Journal of Colloid and Interface Science. 2011;**363**(1):1-24. DOI: 10.1016/j.jcis.2011.07.017

[35] Vermeiren L, Devlieghere F, Debevere J. Effectiveness of some recent antimicrobial packaging concepts. Food Additives and Contaminants Part A. 2002;**19**(Suppl):163-171

[36] Kumar R, Münstedt H. Silver ion release from antimicrobial polyamide/silver composites. Biomaterials. 2005;**26**(14):2081-2088

[37] Necula AM, Dunca S, Stoica I, Olaru N, Olaru L, Ioan S. Morphological properties and antibacterial activity of nano-silver containing cellulose acetate phthalate films. International Journal of Polymer Analysis and Characterization. 2010;**15**(6):341-350. DOI: 10.1080/1023666X.2010.500524

[38] Sánchez Valdes S, Ortega-Ortiz H, Ramos-de Valle LF, Medellín Rodríguez FJ, Guedea MR. Mechanical and antimicrobial properties of multilayer films with a polyethylene/silver nanocomposite layer. Journal of Applied Polymer Science. 2009;**111**(2):953-962. DOI: 10.1002/app.29051

[39] Kim B, Kim D, Cho D, Cho S. Bactericidal effect of TiO2 photocatalyst on selected food-borne pathogenic bacteria. Chemosphere. 2003;**52**(1):277-281

[40] Mariadoss AVA, Ramachandran V, Vijayakumar S, Balupillai A, Herbert FJ, Kumar S, et al. Green synthesis, characterization and antibacterial activity of silver nanoparticles by *Malus domestica* and its cytotoxic effect on (MCF-7) cell line. Microbial Pathogenesis. 2019;**135**:1-9

[41] Aguilar-Méndez MA, Martín-Martínez ES, Ortega-Madan L, Sharma SC, Udayabhanu HR, Suresh D, et al. Facile green fabrication of nanostructure ZnO plates, bullets, flower, prismatic tip, closed pine cone: Their antibacterial, antioxidant, photoluminescent and photocatalytic properties. Spectrochimica Acta Part A: Molecular and Biomolecular Spectroscopy. 2016;**152**:404-416

[42] Rico CM, Majumdar S, Duarte-Gardea M, Peralta-Catalina JR, Marambio-Jones E, Hoek MV. A review of the antibacterial effects of silver nanomaterials and potential implications for human health and the environment. Journal of Nanoparticle Research. 2010;**12**(5):1531-1551. DOI: 10.1007/ s11051-010-9900-y

[43] Gajjar P, Pettee B, Britt DW, Huang W, Johnson WP, Anderson AJ. Journal of Biological Engineering. 2009;**3**:9-10

[44] Baruah S, Dutta J. Nanotechnology applications in pollution sensing and degradation in agriculture: A review. Environmental Chemistry Letters. 2009;**7**:191-204

[45] Ahmed AIS, Raj YD, Su LY. Applications of nickel nanoparticles for control of *Fusarium* wilt on

lettuce and tomato. International Journal of Innovative Research in Science, Engineering and Technology. 2016;**5**(5):7378-7385

[46] Pulit J, Banach M, Szczyg Owska R, Bryk M, et al. Acta Biochimica Polonica. 2013;**60**:795-808

[47] Vahabi K, Mansoori GA, Karimi S. Biosynthesis of silver nanoparticles by fungus *Trichoderma Reesei*. Insciences Journal. 2011;**1**:65-79

[48] Cho JS, Seo YC, Yim TB, Lee HY. Effect of nanoencapsulated vitamin B1 derivative on inhibition of both mycelial growth and spore germination of *Fusarium oxysporum* f. sp. *Raphani*. International Journal of Molecular Sciences. 2013;**14**:4283-4297. DOI: 10.15680/IJIRSET.2016.0505132

[49] Hoglund S. Some electron microscopic studies on the satellite tobacco necrosis virus and its IgG-antibody. The Journal of General Virology. 1968;**2**:427-436

[50] Young M, Willits D, Uchida M, Dougla T. Plant viruses as biotemplates for materials and their use in nanotechnology. Annual Review of Phytopathology. 2008;**46**:361-384

Chapter 5

# Electro-Optical Manipulation Based on Dielectric Nanoparticles

*Jiahao Yan and Yuchao Li*

## Abstract

The ability to dynamically modulate plasmon resonances or Mie resonances is crucial for practical application. Electrical tuning as one of the most efficiently active tuning methods has high switching speed and large modulation depth. Silicon as a typical high refractive index dielectric material can generate strong Mie resonances, which have shown comparable performances with plasmonic nanostructures in spectral tailoring and phase modulation. However, it is still unclear whether the optical response of single silicon nanoantenna can be electrically controlled effectively. In this chapter, we introduce two types of optoelectronic devices based on Mie resonances in silicon nanoantennas. First, we observe obvious blueshift and intensity attenuation of the plasmon-dielectric hybrid resonant peaks when applying bias voltages. Second, photoluminescence (PL) enhancement and modulation are achieved together in the $WS_2$-Mie resonator hybrid system.

**Keywords:** silicon nanoparticles, silicon nanostripes, $WS_2$, active control, photoluminescence manipulation

## 1. Introduction

Dynamically controlling the optical responses from plasmonic or Mie resonators is significant for future optical signal processing [1, 2]. Among different active tuning methods, electrical tuning is one of the most effective one owing to high switching speed and large tuning ranges [3–5]. Recently, electrical tuning on metamaterials based on plasmonic nanostructures has been reported, and the control mechanisms rely on semiconductor layers [6–8], graphene [9–13], or electromechanical deformation [14, 15]. Nevertheless, there are few works about the optoelectronic modulation on nanoscale devices up to now. Furthermore, how to realize the electrical tuning on single nanoparticles is still a challenge.

Combining optical nanoantennas with atomically thin $WS_2$ may be another method to realize dynamic optical responses. Atomically thin $WS_2$ (monolayer or bilayer) exhibits intriguing electrical and optical properties [16–18]. Monolayer $WS_2$ shows strong excitonic emission peak at visible wavelengths; however, ultrathin thickness hinders further enhancement of excitonic emission. Near-field enhancements at excitation wavelengths can enhance the light absorbance, while that at emission wavelengths would boost the emission rate, so those two factors both enhance the excitonic emission from $WS_2$. Based on it, many efforts have been made to realize field enhancements via plasmonic nanostructures and photonic crystals at both excitation and emission wavelengths.

Silicon nanoantennas as a typical dielectric Mie resonator have wide application prospect in building metasurfaces [19–21], nonlinear optics [22], and biosensing [23]. They may be better choice than plasmonic structures and photonic crystals in building electrically controlled devices. The Mie resonances in silicon nanoantennas can be modulated through changing the sizes [24, 25] or crystallographic phases [20] passively. However, how to realize active control based on the Mie resonances in silicon nanocavities is still a challenge. Besides changing the optical properties of Mie resonators intrinsically, active tuning may also be realized via coupling with 2D materials. Neshev et al. have theoretically demonstrated the PL modulation of 2D materials based on the directional emission caused by Mie resonators [26]. Recently, the first experimental work has been done, in which a forward-to-backward emission ratio of 20 was realized because of the interaction between $MoS_2$ monolayer and Mie resonators [27]. However, both of them were analyzed on passive control.

# 2. Electrically controlled optical responses of silicon-based nanoantennas

In this chapter, we will discuss the applications of silicon-based Mie resonators into electro-optical modulation. This chapter can be divided into two parts:

First, we demonstrated the electrically tunable scattering of a single silicon nanoparticle at visible wavelengths. To build the nanoantennas, gold interdigital electrodes with separation distances between 100 and 200 nm were fabricated using photolithography and focused ion beam (FIB) milling. After trapping silicon nanoparticles with different sizes between adjacent two electrodes, the scattering spectra under different voltages can be measured. Interestingly, the scattering experiences blueshift and obvious intensity attenuation when increasing the applied voltages from 0 to 1.5 V. In theory, MIS (metal-insulator-semiconductor) junctions can be formed at Au-$SiO_2$-Si interfaces [28]. Once the bias voltage increases, the inversion and accumulation effect would produce much more free carriers at interfaces [29–31] and then change the permittivity based on Drude model [32, 33]. The proposed hybrid nanoantennas represent a new method to build optoelectronic devices based on Mie resonators.

Second, we combined silicon nanostripes, a typical Mie resonator, with $WS_2$ to realize active PL manipulation. In the proposed electro-optical modulator, suspended monolayer and bilayer $WS_2$ are covered on a Si nanostripe. The Si nanostripe not only acts as a nanoscale gate electrode but also a Mie resonator. For both monolayer and bilayer $WS_2$, the PL intensities on the nanostripes are much stronger than those of the suspended one. After applying gate voltages, both the electrostatic doping and strain come into effect. This new tuning mechanism leads to abnormal control of exciton emission from $WS_2$, which is clearly different from that in previous works [34–36]. Considerable PL tuning can also be observed in bilayer $WS_2$ gated by Si nanostripes. Based on the modulation capability, we believe the proposed electro-optical modulator will bring new possibilities for future nanophotonic devices.

# 3. Hybrid nanoantennas based on silicon nanoparticles and nanostripes

## 3.1 Electrically driven scattering in a hybrid dielectric-plasmonic nanoantenna

In order to build an electrically controlled silicon nanoantenna, the biggest issue is how to apply voltage on a single nanoparticle and collect the electrically modulated signals with low noise. The design of electrically tunable silicon nanoantenna is shown in **Figure 1a**. First, maskless laser lithography and electron-beam deposition were used to fabricate Au electrodes with the thickness of 100 nm on the Si/$SiO_2$ substrate, and the thickness of $SiO_2$ layer is 300 nm. In our design, several large Au electrodes (200 × 400 μm) are deposited with a row of holes in the center. Second, the connected area in the center was nano-patterned using FIB milling to form nanoscale interdigital electrode structure. The separation distance between adjacent nano-electrodes is adjusted from 100 to 200 nm to match the size distributions of silicon nanoparticles, since the silicon nanoparticles fabricated through femtosecond laser ablation in liquid (fs-LAL) have a wide size distribution. Finally, during the evaporation process, the silicon nanoparticles in colloid have a certain probability to be trapped in the gaps.

Before studying the optical properties of Si-Au hybrid nanoantennas, we should study the Au electrode platform first. For the fabricated Au grating, due to the incident light that comes from a dark-field circle in the objective, wave vectors with different directions at x-y plane cannot launch surface plasmon polariton efficiently. In addition, the plasmon energy mainly decays nonradiatively through near-field coupling between adjacent Au electrodes, so Au gratings cannot show bright scattering as shown in **Figure 1b**. However, if only two electrodes left (see **Figure 1b**), localized surface plasmon can be formed between two Au electrodes. Strong scattering light can be generated from the plasmonic field enhancement in the gap. Therefore, we use Au grating in experiment whose scattering can be ignored compared with Si nanoparticles. Typical Au electrode-loaded Si nanoparticles are shown in **Figure 1c** and **d**, where a bright dot can be seen in dark-field image which means the scattering from the Si nanoantenna. In spectral measurement, through moving the scattering spot into the center of slit and only extracting the data from the location of nanoparticle, the exact scattering from the Si nanoparticle can be obtained.

For isolated Si nanoparticles, the resonant modes depend on particle sizes and particle numbers according to Mie theory. While for Au electrode-loaded Si

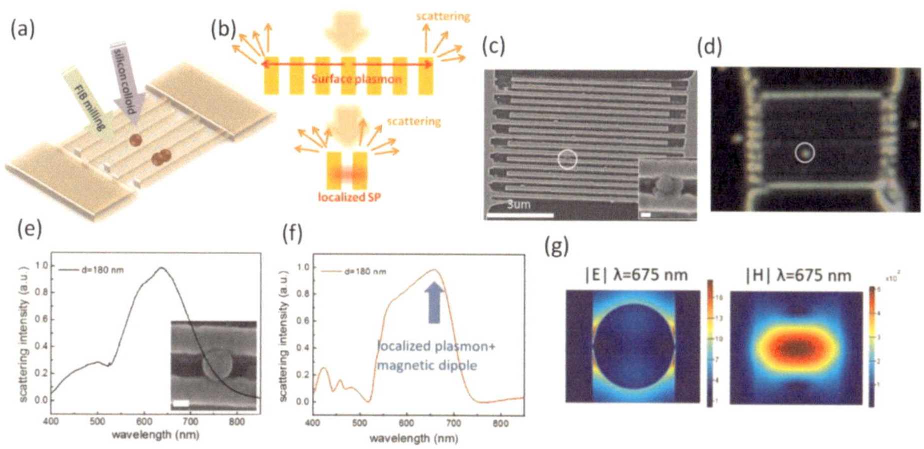


**Figure 1.**
*Optical properties of the silicon nanoantenna. (a) A schematic diagram explains the fabrication of Au electrode-loaded Si nanoparticles. (b) The schematic shows different plasmon resonant modes of two types Au electrodes. (c) The scanning electron microscope (SEM) image of Au interdigital electrodes with a silicon NP trapped among them. Inset is the high magnification SEM image with a scale bar of 100 nm. (d) The dark-field scattering image of the sample in (c). The white circle reveals the location of Si nanoparticles. (e, f) Measured scattering spectrum of a 180 nm Si nanoparticle (e) and the corresponding simulated scattering spectrum (f). (g) The electric and magnetic field distributions at 675 nm, which represent the hybrid modes coupling between localized plasmon and magnetic dipole.*

nanoparticles, the mode coupling between nanoparticles and Au electrodes also needs to be considered. The hybrid nanoantennas may exhibit different scattering spectra at visible wavelengths. Therefore, it is necessary to study the scattering spectra without applied voltage first. Although self-assembled process is random, desirable and representative nanoparticles can be found through matching and positioning. **Figure 1e** shows 180 nm Si nanoparticles between two Au electrodes with a spacing slightly less than 180 nm. The measured scattering spectra exhibits a single broad peak around $\lambda = 650\,nm$. As shown in **Figure 1f**, the simulated spectrum is very similar to experimental spectrum. Corresponding electric and magnetic field distributions in **Figure 1g** demonstrate the existence of circular magnetic field distributions and strong electric field enhancement at interfaces, which means the scattering peak is generated from the interaction between the Mie-type magnetic dipole mode in Si nanoparticles and the localized surface plasmon resonances (LSPR) at Au-Si interfaces.

The electrical properties of the Si-Au hybrid devices were measured using a semiconductor parameter analyzer. The measured I-V curve is shown in **Figure 2a**, and we can conclude that the Si-Au interfaces can be regarded as Schottky junctions. From 0 to 1.5 V, the current increases nonlinearly with the voltage. For the fabricated Si nanoparticles, thin oxide (1–2 nm) shells will be formed inevitably in the air. Therefore, the interfaces are MIS junctions whose current is generated through tunnel effect and plasmon hot electron injection. For MIS junctions, the band bends upward at interfaces when no voltage applies as the schematic diagram shown in **Figure 2b**. Depletion region forms at the interfaces and free carriers move away from interfaces based on the band bending [29]. With applied bias, surface potential at two interfaces increases. The carrier concentration at the MIS junction under lower potential was greatly increased because the downward energy band realizes the accumulation of electrons. The other MIS junction under higher potential could form an inversion layer if the applied voltage is high enough. When the intrinsic energy level crosses the Fermi level [29–31], the hole density would greatly increase under the inversion state. The charge densities at surface at different applied voltages can be estimated by the following equations [29]

$$|Q_{inv}| = -C_{ox}(V_G - V_T) \propto \exp\left(\frac{\beta\Psi_S}{2}\right) \quad (1)$$

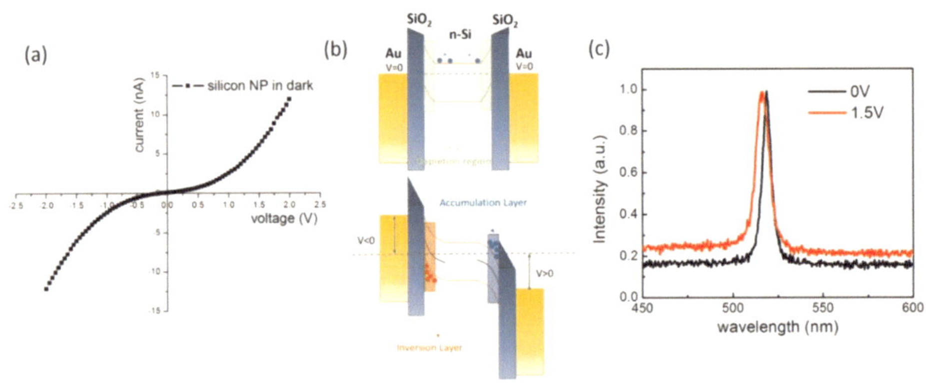


**Figure 2.**
*Analysis on the voltage-induced carrier injection. (a) The I-V curve of a fabricated Si-Au hybrid structure. (b) Schemes for the band bending and carrier distribution with and without applied voltage. (c) The Raman spectrum of a loaded nanoparticle before and after applied voltage (1.5 V).*

$$|Q_{acc}| = C_{ox}(V_G - V_{FB}) \propto \exp\left(\frac{-\beta\Psi_S}{2}\right) \tag{2}$$

where $Q_{acc}$ and $Q_{inv}$ are the carrier densities at accumulation and inversion regions. $C_{ox}$ is the capacitance of thin insulator layer. $V_G$ is the applied bias. $V_{FB}$ and $V_T$ are accumulation and inversion effects related to voltages. $\beta = q/k_BT$ is a constant. $\Psi_S$ is the surface potential at interfaces. Based on the electric field we applied and above calculation, the electron or hole concentrations can be increased by more than three orders of magnitude in accumulation or inversion regions, respectively [29]. The Raman signals under different voltages are presented in **Figure 2c**. With 1.5 V applied bias, the resonant peak of silicon just red shifts slightly, and this weak shift means the temperature variation is less than 100 K [37]. Therefore, we can exclude the influence of thermal effect on the refractive index since the refractive index of silicon only increases $3.85 \times 10^{-4}$ per degree [38].

To examine whether the mechanism discussed above could affect the optical properties significantly, the scattering spectra of the typical silicon nanoantenna with applied bias were presented in **Figure 3a**. To ensure stability, all scattering data were collected in 1 min during the increase of voltage from 0 to 1.5 V. Because the interfaces of the fabricated hybrid nanoantenna are symmetric, we only need to collect the scattering spectra under forward bias which is enough to embody the properties of hybrid nanoantennas. For a typical hybrid nanoantenna as shown in **Figure 3a** with a 180 nm Si nanoparticle, we can observe the suppression of hybrid plasmon-Mie resonant peaks when increasing the voltages. The magnetic dipole peak was dominated when no voltage applies. However, when applied voltage reaches 1.5 V, the electric dipole peak at shorter wavelength becomes the more prominent one.

As discussed above, different applied voltages result in different free carrier concentrations of Si nanoparticles. Further, we should clarify how carrier injection influences the dielectric function of silicon. The modulation mechanism is based on free carrier-induced refractive index change. Although electric field cannot change the refractive index of bulk silicon or whole silicon nanostructures significantly as previous works reported [39], obvious refractive index modification can be realized at accumulation and inversion interfaces. From the field profiles, one can understand the refractive index change on surface is enough to change optical responses because field enhancements and radiative decays mainly come from interfaces. How free carriers contribute to the refractive index change at interfaces can be described by the Drude model

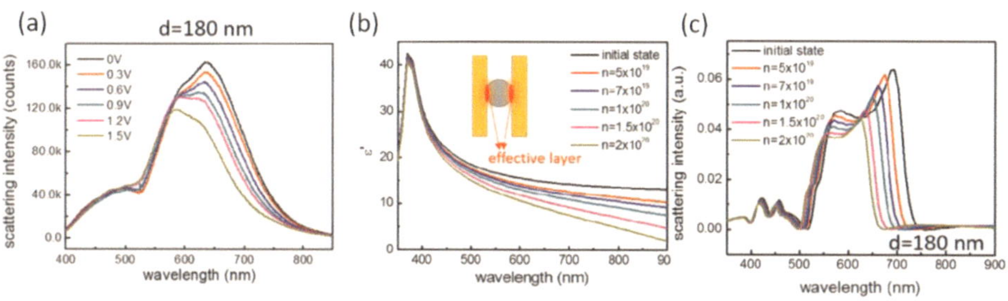


**Figure 3.**
*Electrically controlled scattering. (a) Scattering spectra of the 180 nm Si nanoparticles when applied voltages equal to 0, 0.3, 0.6, 0.9, 1.2, and 1.5 V. (b) The variation trend of the real part of permittivity at Au-$SiO_2$-Si interfaces when increasing the carrier concentrations. (c) The calculated scattering spectra of the 180 nm Si nanoparticle under different carrier concentrations at interfaces.*

$$\Delta\varepsilon' = -\frac{\varepsilon_\infty \omega_p^2 \tau^2}{1+\omega^2\tau^2} \tag{3}$$

$$\Delta\varepsilon'' = \frac{\varepsilon_\infty \omega_p^2 \tau}{\omega(1+\omega^2\tau^2)} \tag{4}$$

where $\omega_p$ is the plasma frequency which is defined as $\omega_p = \sqrt{Ne^2/m_C\varepsilon_\infty\varepsilon_0}$. $\tau$ is the damping time equals to $\mu m_C/e$ where $e$ is the charge of an electron. $m_C$ is the effective mass, $\varepsilon_0$ is the vacuum permittivity, and $\varepsilon_\infty$ is the permittivity of silicon at visible band. $N$ is the concentration of free carrier which determines the changes of permittivity. Using the Drude model discussed above, we can calculate how free carriers influence the dielectric function of silicon as shown in **Figure 3b**. Putting different carrier concentrations ($10^{17}$to$2.0 \times 10^{20}$cm$^{-3}$) into Drude model, one can see the real part of permittivity decreases gradually especially at longer wavelengths from 600 to 900 nm. Because the accumulation and inversion layers are less than 5 nm at interfaces, we only used the free carrier-induced dielectric functions at interfaces for the numerical simulation. As shown in **Figure 3c**, the simulated scattering spectra under different carrier concentrations are very similar to the corresponding measured spectra under different applied bias. For the 180 nm Si nanoparticle (see **Figure 3c**), the hybrid resonant peak experiences blueshift and intensity attenuation when increasing the carrier concentrations in sequence. The attenuation trend of resonant peaks is very similar to the experimental spectra in **Figure 3a**. Our proposed structures provide an opportunity to collect the electrically controlled scattering signals on single-particle level.

## 3.2 PL enhancements enabled by silicon nanostripes

Owing to the unique properties of dielectric Mie resonators, researchers are trying to use Mie resonators as an important building block to form new-generation electro-optical modulators. One strategy is to combine Mie resonators with 2D materials as the schematic shown in **Figure 4a**. $WS_2$ monolayers and bilayers were obtained by mechanical exfoliation and all-dry transfer technique. $WS_2$ layers and Si nanostripes were aligned and contacted under an optical microscope. Si nanostripes were fabricated by FIB milling onto $SiO_2$ coated silicon-on-insulator (SOI) wafers. Gold electrodes were patterned and deposited on $WS_2$ and bare silicon to build source, drain, and gate. A simple cross-section schematic of the substrate in **Figure 4a** shows that there is an insulator layer between the Si substrate and the top Si film. Therefore, the scattering from Si nanostripes is not only pure Mie effect but the Mie resonance combined with the Fabry-Perot effect. In our case, the thickness of the insulator layer (h) is 375 nm. The dark-field scattering spectrum and the corresponding optical image in **Figure 1b** indicate that Si nanostripes have a broadband resonant peak. Dominant peaks are located around 700 nm, and two small peaks can also be distinguished below $\lambda = 600$ nm. **Figure 4c** is a typical $WS_2$-Si nanostripe hybrid nanostructure. Corresponding SEM images are shown in **Figure 4d**. From SEM images, we can see the width of Si nanostripe is around 650 nm surrounded by two $10 \times 30\,\mu m$ etched regions. Wrinkles and missing regions are inevitably formed during the transfer and lift-off processes. Fortunately, these regions can be avoided in the following measurements.

**Figure 5a** indicates different locations we measured on bilayer and monolayer $WS_2$. From the PL emission spectra of monolayer $WS_2$ as shown in **Figure 5b**, one can conclude Si nanostripes can increase the PL intensities but less than threefold compared with that in the suspended region. It should be noticed that the

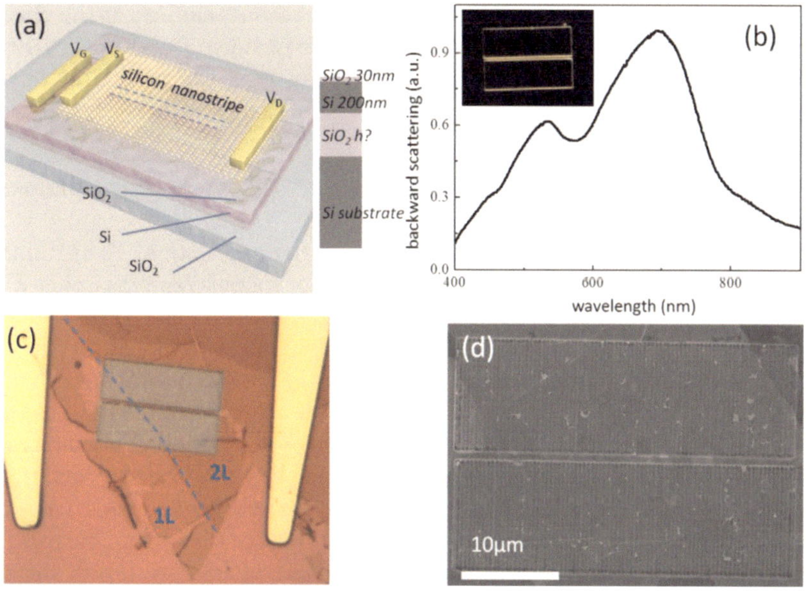


**Figure 4.**
*Experimental design for the $WS_2$-Si nanostripe hybrid structure. (a) Schematic illustration of the electrically controlled device and the cross section of the $SiO_2$-coated SOI substrate. (b) Dark-field backward scattering of the fabricated Si nanostripe. Inset: The dark-field optical image. (c) The bright-field optical image of a typical device. Monolayer and bilayer regions are labeled as 1L and 2L. (d) The corresponding SEM image.*

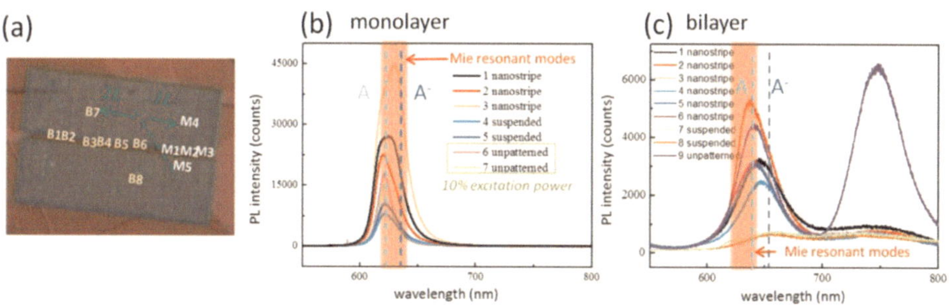


**Figure 5.**
*PL enhancements in monolayer and bilayer $WS_2$. (a) The optical image showing the detection points on monolayer (M) and bilayer (B) $WS_2$. (b, c) PL spectra of different positions marked in (a). The locations of exciton (A) and trion ($A^-$) states are labeled, along with the range of Mie resonant modes (marked by red stripe).*

unpatterned region can also enhance the PL intensity and the enhancement performance is better than that on the Si nanostripe. For bilayer $WS_2$, the PL enhancement is much more significant as shown in **Figure 5c**. The bilayer $WS_2$ on the Si nanostripe possesses nearly 10 times larger PL intensities than the suspended area. Interestingly, this PL enhancement was only observed at $\lambda = 625\,nm$ where direct bandgap transition happens. However, for unpatterned area, PL intensities at both direct and indirect transition wavelengths ($\lambda = 635\&750$ nm) are enhanced. The performances of PL enhancement for direct transition are comparable for unpatterned area and the Si nanostripe. Moreover, unpatterned area can strengthen the indirect transition more than 10 times. Besides the differences of PL intensities, the line shapes also change in **Figure 5b** and **c** which reveals the conversion between exciton (A) and trion ($A^-$). The locations of exciton and trion emission are labeled in **Figure 5b** and **c**. Further peak fitting demonstrates the decreased trend of trion emission from suspended area and nanostripes to the unpatterned region. Substrate effect and the optical resonant modes may be two reasons that lead to the change of line shapes. The Fabry-Perot mode in the unpatterned substrate and Fabry-Perot mode assisted by Mie resonances in fabricated nanostripes can both influence the PL line shapes.

### 3.3 PL manipulation of monolayer and bilayer $WS_2$ gated by silicon nanostripes

Realizing the electrical tuning is crucial for further application. To examine the electrical tuning performance, first, PL intensities of monolayer $WS_2$ under different voltages were measured as shown in **Figure 6a**. When applying negative gate voltages from 0 to −10 V, the maximum PL intensity increases by 50%. On the contrary, the intensity of the PL peak decreases to half under positive gate voltages from 0 to 10 V. Compared with the normal $WS_2$ monolayer gated by the flat gate [34–36], the PL enhancement effect is weaker, while the reduction effect is more obvious. This phenomenon indicates that the tuning effect is not pure electrostatic doping and there should be a new mechanism. PL changes of the bilayer $WS_2$ under different voltages were also measured as shown in **Figure 6b** and **c**. Unexpectedly, the variation trends under positive and negative gate voltages are almost the same. The maximum PL intensity doubled when increasing the gate voltage from 0 to 10 V. Several groups have studied the electrically controlled PL of bilayer $WS_2$, while no obvious effect has been observed [34–36]. Therefore, the obvious PL enhancement we observed may not arise from pure electrostatic doping. The gate voltage dependent intensity of excitonic peak is plotted in **Figure 6d**. The PL intensity of monolayer $WS_2$ increases linearly with gate voltage, while the PL intensity of bilayer $WS_2$ and the gate voltage follow a parabolic relationship.

How to explain the abnormal PL manipulation in the proposed hybrid nanoantennas? The schematic shown in **Figure 7a** may give a better understanding.

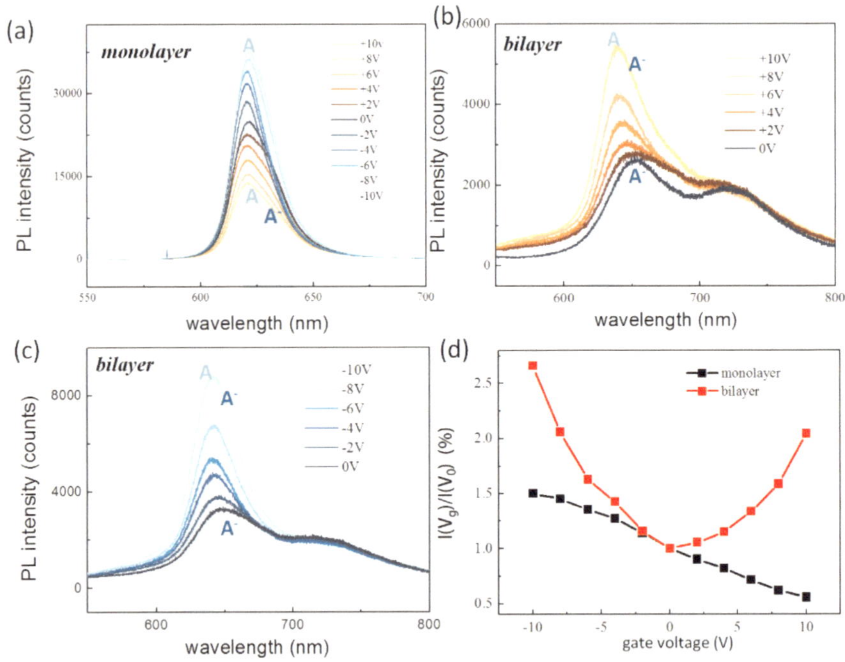


**Figure 6.**
*Electrically controlled PL. (a) PL spectra of the monolayer $WS_2$ on the Si nanostripe at different gate voltages. (b, c) PL spectra of the bilayer $WS_2$ on the Si nanostripe at different gate voltages. (d) The gate voltage dependence of PL intensities.*

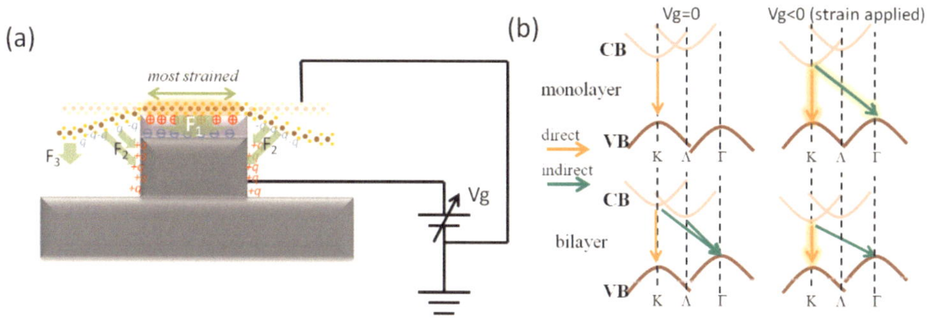


**Figure 7.**
*Mechanism of electrical tuning. (a) Schematic setup showing how tensile strain can be generated by electrostatic gating. $F_1$, $F_2$, and $F_3$ represent three types of electrostatic forces. (b) Schematic diagram showing the changes of band structure and dominated transitions before and after applying strain.*

When placing $WS_2$ flakes on the Si nanostripe, there are three regions that experience different forces. The first part is $WS_2$ on the Si nanostripe, the second part is $WS_2$ near the edges of Si nanostripes, and the third part is fully suspended $WS_2$. If the applied voltage is high enough, a great number of holes and electrons will be produced at $WS_2$ layers and bottom Si nanostructures, respectively. Besides the electrostatic doping, the static electric field can also produce attractive forces. As shown in **Figure 7a**, there are three types of attractive forces $F_1$, $F_2$, and $F_3$ which depend on different distances and capacitances. $F_1$ is the attractive force between adherent $WS_2$ and Si nanostripe through the 30 nm insulator layer. $F_2$ is the attractive force between suspended $WS_2$ around edges and the Si nanostripe. $F_2$ at edges equals to $F_1$ and decreases gradually away from the Si nanostripe, so this force will let $WS_2$ at edges be curved and exert a large uniaxial tensile strain on $WS_2$. The electrostatic attraction between the suspended $WS_2$ and bottom Si ($F_3$) is much weaker which can be ignored because of a larger distance and smaller capacitance. The deflection of few layer $WS_2$ $\Delta l$ under electrostatic force can be calculated by equation [40]:

$$PL^2 = 8T_0 t\Delta l + \frac{64}{3}\frac{Et}{L^2(1-\upsilon^2)}\Delta l^3 \tag{5}$$

where $t$ is the thickness that equals to 0.8 or 1.6 nm, $E$ is the Young's modulus of $WS_2$, and $L$ is the effective length of strained $WS_2$. The effective length is very small because only $WS_2$ at edges experiences significant attractive forces. $P = \left(C^2V_g^2\right)/2\varepsilon_0$ is the electrostatic pressure, where $C$ is the capacitance per unit area and $\varepsilon_0$ is the permittivity of vacuum. After considering the relationship between strain and the deflection, the strain can be estimated by [40]:

$$\chi = 2\Delta l^2/L^2 = 2[(3/64)(PL)/(Et)]^{2/3} \tag{6}$$

Based on the calculation above, the strain $\chi$ within effective area is larger than 2.8%. Such high strain is able to change the band structure of $WS_2$ and influence the excitonic emission [41–43]. Therefore, in **Figure 7b**, we combine electrostatic doping and strain effect together to analyze the change of band structure and PL intensity. For 1L-$WS_2$, if only electrostatic doping comes into effect, negative gating would enhance the PL intensity contributed by direct transition. However, in our case, larger strain generated from attractive forces changes the band gap of monolayer $WS_2$. Without applied bias, the locations of valence-band maximum and conduction-band minimum are overlapped at the point K. Once strain is applied, the valence-band maximum will shift from K to Γ point, and indirect transition will happen. Therefore, the strain effect will weaken the PL intensity, which has opposite effect compared with electrostatic doping under negative gating. As a consequence, the PL enhancement is weaker under negative gating and more obvious under positive gating compared with previous works. For 2L-$WS_2$, electrostatic effect can be ignored, and the strain effect is dominated. Without applied bias, the PL emission of bilayer $WS_2$ contains both direct transitions and indirect transitions. If the strain becomes larger, valence-band maximum at the K point reduces, which promotes the direct transition along the $K - K$ direction [43, 44].

## 4. Dielectric nanoparticles for bionanosensing

From the above analysis, we know that Mie resonators such as Si nanoparticles can combine with plasmonic nano-electrodes to obtain electrically controlled optical responses, and Mie resonators such as Si nanostripes can also interact with $WS_2$ layers to realize abnormal electro-optical modulation based on electrostatic doping and strain effect. Further, it is necessary to utilize the unique properties of Mie resonators and analyze their application prospect in biosensing.

As we know, plasmonic nanostructures have been widely used in biosensing. Plasmon resonances experience redshift when increasing the surrounding refractive index, which is the most basic mechanism of biosensing. Dielectric Mie resonators have low-loss feature and strong directional scattering which also have a potential as biosensing nanoantennas. However, based on current reports and our experiments, we found the optical responses of single silicon nanostructures such as Si nanoparticles cannot exhibit obvious change when changing the surrounding refractive index. Therefore, the biosensor based on a single Si nanoparticle is insensitive.

Fortunately, we found the scattering spectra become very sensitive to surrounding refractive index if single Si nanoparticles combine to dimers or other oligomers. Based on our theoretical analysis, touching Si nanoparticles can produce strong electric field enhancement in the gap. This gap electric mode is a key factor for sensitive spectral change, because the gap electric mode would enhance and experience redshift with the increase of surrounding refractive index. As talked above, 1–2 nm silica layer is naturally grown on Si nanoparticles. Based on the mature biomarker technique, we can easily modify the silica surface with specific functional groups and realize the detection of many kinds of biomolecules. Furtherly, Si nanoparticles can be injected into living cells to realize the sensing in vivo. Finally, we can combine the biosensing and optoelectronic property of Mie resonators to build new type biosensors. On the one hand, biomolecules can change the electrical properties of dielectric nanostructures and then influence the optical signals. On the other hand, biomolecules can change the optical properties of nanoantennas and furtherly influence the electrical readout.

## 5. Conclusions

In this chapter, we have introduced the electrically controlled scattering of individual Mie resonators and PL from the $WS_2$-Mie resonator hybrid system. The strong magnetic responses and low-loss feature make silicon-based Mie resonators become important building blocks in nanophotonics. Combining top-down and bottom-up fabrication methods, plasmon-Mie hybrid nanostructures and $WS_2$-Mie hybrid nanostructures are fabricated, respectively. These structures give us an opportunity to apply voltages at nanoscale and collect the optical signals at single points. Interfaces are important in those hybrid nanodevices. The interfaces between plasmonic structures and Mie resonators bring new mechanism on carrier injection and changes of refractive index, while the contact between $WS_2$ and Mie resonator generates unique PL active tuning arising from the synergistic effect between electrical doping and tensile strain under gate voltages. In the emerging applications based on dielectric Mie resonators, our findings provide an important and feasible method to build optoelectronic functional devices that can transfer electrical signal to optical signal. Furthermore, the excellent biosensing performance will expand the applications of Mie resonator-based optoelectronic devices.

## Acknowledgements

This work was supported by the National Natural Science Foundation of China (No. 11774135, 11874183, and 61827822).

## Conflict of interest

The authors declare no competing financial interests.

## Author details

Jiahao Yan and Yuchao Li*
Institute of Nanophotonics, Jinan University, Guangzhou, China

*Address all correspondence to: liyuchao@jnu.edu.cn

## References

[1] Lee SH, Choi M, Kim T-T, Lee S, Liu M, Yin X, et al. Switching terahertz waves with gate-controlled active graphene metamaterials. Nature Materials. 2012;**11**:936-941. DOI: 10.1038/nmat3433

[2] Kang L, Lan S, Cui Y, Rodrigues SP, Liu Y, Werner DH, et al. An active metamaterial platform for chiral responsive optoelectronics. Advanced Materials. 2015;**27**:4377-4383. DOI: 10.1002/adma.201501930

[3] Kern J, Kullock R, Prangsma J, Emmerling M, Kamp M, Hecht B. Electrically driven optical antennas. Nature Photonics. 2015;**9**:582-586. DOI: 10.1038/nphoton.2015.141

[4] Fan P, Yu Z, Fan S, Brongersma ML. Optical Fano resonance of an individual semiconductor nanostructure. Nature Materials. 2014;**13**:471-475. DOI: 10.1038/nmat3927

[5] Prangsma JC, Kern J, Knapp AG, Grossmann S, Emmerling M, Kamp M, et al. Electrically connected resonant optical antennas. Nano Letters. 2012;**12**:3915-3919. DOI: 10.1021/nl3007374

[6] Chen H-T, Padilla WJ, Zide JM, Gossard AC, Taylor AJ, Averitt RD. Active terahertz metamaterial devices. Nature. 2006;**444**:597-600. DOI: 10.1038/nature05343

[7] Chen H-T, Lu H, Azad AK, Averitt RD, Gossard AC, Trugman SA, et al. Electronic control of extraordinary terahertz transmission through subwavelength metal hole arrays. Optics Express. 2008;**16**:7641-7648. DOI: 10.1364/OE.16.007641

[8] Jun YC, Gonzales E, Reno JL, Shaner EA, Gabbay A, Brener I. Active tuning of mid-infrared metamaterials by electrical control of carrier densities. Optics Express. 2012;**20**:1903-1911. DOI: 10.1364/OE.20.001903

[9] Mousavi SH, Kholmanov I, Alici KB, Purtseladze D, Arju N, Tatar K, et al. Inductive tuning of Fano-resonant metasurfaces using plasmonic response of graphene in the mid-infrared. Nano Letters. 2013;**13**:1111-1117. DOI: 10.1021/nl304476b

[10] Emani NK, Chung T-F, Kildishev AV, Shalaev VM, Chen YP, Boltasseva A. Electrical modulation of fano resonance in plasmonic nanostructures using graphene. Nano Letters. 2013;**14**:78-82. DOI: 10.1021/nl403253c

[11] Yao Y, Shankar R, Kats MA, Song Y, Kong J, Loncar M, et al. Electrically tunable metasurface perfect absorbers for ultrathin mid-infrared optical modulators. Nano Letters. 2014;**14**:6526-6532. DOI: 10.1021/nl503104n

[12] Yao Y, Kats MA, Genevet P, Yu N, Song Y, Kong J, et al. Broad electrical tuning of graphene-loaded plasmonic antennas. Nano Letters. 2013;**13**:1257-1264. DOI: 10.1021/nl3047943

[13] Valmorra F, Scalari G, Maissen C, Fu W, Schönenberger C, Choi JW, et al. Low-bias active control of terahertz waves by coupling large-area CVD graphene to a terahertz metamaterial. Nano Letters. 2013;**13**:3193-3198. DOI: 10.1021/nl4012547

[14] Ou J-Y, Plum E, Zhang J, Zheludev NI. An electromechanically reconfigurable plasmonic metamaterial operating in the near-infrared. Nature Nanotechnology. 2013;**8**:252-255. DOI: 10.1038/nnano.2013.25

[15] Chen K, Razinskas G, Feichtner T, Grossmann S, Christiansen S, Hecht B. Electromechanically tunable suspended optical nanoantenna. Nano Letters.

2016;**16**:2680-2685. DOI: 10.1021/acs.nanolett.6b00323

[16] Wang QH, Kalantar-Zadeh K, Kis A, Coleman JN, Strano MS. Electronics and optoelectronics of two-dimensional transition metal dichalcogenides. Nature Nanotechnology. 2012;**7**:699-712. DOI: 10.1038/nnano.2012.193

[17] Zhao W, Ghorannevis Z, Chu L, Toh M, Kloc C, Tan P-H, et al. Evolution of electronic structure in atomically thin sheets of $WS_2$ and $WSe_2$. ACS Nano. 2012;**7**:791-797. DOI: 10.1021/nn305275h

[18] Fiori G, Bonaccorso F, Iannaccone G, Palacios T, Neumaier D, Seabaugh A, et al. Electronics based on two-dimensional materials. Nature Nanotechnology. 2014;**9**:768-779. DOI: 10.1038/nnano.2014.207

[19] Li Y, Kita S, Muñoz P, Reshef O, Vulis DI, Yin M, et al. On-chip zero-index metamaterials. Nature Photonics. 2015;**9**:738-742. DOI: 10.1038/nphoton.2015.198

[20] Zywietz U, Evlyukhin AB, Reinhardt C, Chichkov BN. Laser printing of silicon nanoparticles with resonant optical electric and magnetic responses. Nature Communications. 2014;**5**:3402. DOI: 10.1038/ncomms4402

[21] Arbabi A, Horie Y, Bagheri M, Faraon A. Dielectric metasurfaces for complete control of phase and polarization with subwavelength spatial resolution and high transmission. Nature Nanotechnology. 2015;**10**:937-943. DOI: 10.1038/nnano.2015.186

[22] Ma C, Yan J, Liu P, Wei Y, Yang G. Second harmonic generation from an individual all-dielectric nanoparticle: Resonance enhancement versus particle geometry. Journal of Materials Chemistry C. 2016;**4**:6063-6069. DOI: 10.1039/C6TC01635A

[23] Yan J, Liu P, Lin Z, Yang G. New type high-index dielectric nanosensors based on the scattering intensity shift. Nanoscale. 2016;**8**:5996-6007. DOI: 10.1039/C5NR07871G

[24] Yan J, Lin Z, Liu P, Yang G. A design of Si-based nanoplasmonic structure as an antenna and reception amplifier for visible light communication. Journal of Applied Physics. 2014;**116**:154307. DOI: 10.1063/1.4898684

[25] Yan J, Ma C, Liu P, Wang C, Yang G. Generating scattering dark states through the fano interference between excitons and an individual silicon nanogroove. Light: Science & Applications. 2017;**6**:e16197. DOI: 10.1038/lsa.2016.197

[26] Chen H, Nanz S, Abass A, Yan J, Gao T, Choi D-Y, et al. Enhanced directional emission from monolayer $WSe_2$ integrated onto a multiresonant silicon-based photonic structure. ACS Photonics. 2017;**4**:3031-3038. DOI: 10.1021/acsphotonics.7b00550

[27] Cihan AF, Curto AG, Raza S, Kik PG, Brongersma ML. Silicon Mie resonators for highly directional light emission from monolayer $MoS_2$. Nature Photonics. 2018;**12**:284-290. DOI: 10.1038/s41566-018-0155-y

[28] Léonard F, Talin AA, Swartzentruber B, Picraux S. Diameter-dependent electronic transport properties of Au-catalyst/Ge-nanowire Schottky diodes. Physical Review Letters. 2009;**102**:106805. DOI: 10.1103/PhysRevLett.102.106805

[29] Sze SM, Ng KK. Physics of Semiconductor Devices. USA: John Wiley & Sons; 2006

[30] Hauser J, Littlejohn M. Approximations for accumulation and inversion space-charge layers in semiconductors. Solid State Electronics.

1968;**11**:667-674. DOI: 10.1016/0038-1101(68)90069-5

[31] Qian F, Kim DM, Kawamoto GH. Inversion/accumulation-mode polysilicon thin-film transistors: Characterization and unified modeling. IEEE Transactions on Electron Devices. 1988;**35**:1501-1509. DOI: 10.1109/16.2583

[32] Lewi T, Iyer PP, Butakov NA, Mikhailovsky AA, Schuller JA. Widely tunable infrared antennas using free carrier refraction. Nano Letters. 2015;**15**:8188-8193. DOI: 10.1021/acs.nanolett.5b03679

[33] Sokolowski-Tinten K, von der Linde D. Generation of dense electron-hole plasmas in silicon. Physical Review B. 2000;**61**:2643. DOI: 10.1103/PhysRevB.61.2643

[34] Mak KF, He K, Lee C, Lee GH, Hone J, Heinz TF, et al. Tightly bound trions in monolayer $MoS_2$. Nature Materials. 2013;**12**:207-211. DOI: 10.1038/nmat3505

[35] Newaz A, Prasai D, Ziegler J, Caudel D, Robinson S, Haglund R Jr, et al. Electrical control of optical properties of monolayer $MoS_2$. Solid State Communications. 2013;**155**:49-52. DOI: 10.1016/j.ssc.2012.11.010

[36] Pei J, Yang J, Xu R, Zeng YH, Myint YW, Zhang S, et al. Exciton and trion dynamics in bilayer $MoS_2$. Small. 2015;**11**:6384-6390. DOI: 10.1002/smll.201501949

[37] Doerk GS, Carraro C, Maboudian R. Temperature dependence of Raman spectra for individual silicon nanowires. Physical Review B. 2009;**80**:073306. DOI: 10.1103/PhysRevB.80.073306

[38] Hoyland J, Sands D. Temperature dependent refractive index of amorphous silicon determined by time-resolved reflectivity during low fluence excimer laser heating. Journal of Applied Physics. 2006;**99**:063516. DOI: 10.1063/1.2186378

[39] Xu Q, Schmidt B, Pradhan S, Lipson M. Micrometre-scale silicon electro-optic modulator. Nature. 2005;**435**:325-327. DOI: 10.1038/nature03569

[40] Bao W, Myhro K, Zhao Z, Chen Z, Jang W, Jing L, et al. In situ observation of electrostatic and thermal manipulation of suspended graphene membranes. Nano Letters. 2012;**12**:5470-5474. DOI: 10.1021/nl301836q

[41] Desai SB, Seol G, Kang JS, Fang H, Battaglia C, Kapadia R, et al. Strain-induced indirect to direct bandgap transition in multilayer $WSe_2$. Nano Letters. 2014;**14**:4592-4597. DOI: 10.1021/nl501638a

[42] Hui YY, Liu X, Jie W, Chan NY, Hao J, Hsu Y-T, et al. Exceptional tunability of band energy in a compressively strained trilayer $MoS_2$ sheet. ACS Nano. 2013;7:7126-7131. DOI: 10.1021/nn4024834

[43] Dhakal KP, Roy S, Jang H, Chen X, Yun WS, Kim H, et al. Local strain induced band gap modulation and photoluminescence enhancement of multilayer transition metal dichalcogenides. Chemistry of Materials. 2017;**29**:5124-5133. DOI: 10.1021/acs.chemmater.7b00453

[44] Roldán R, Castellanos-Gomez A, Cappelluti E, Guinea F. Strain engineering in semiconducting two-dimensional crystals. Journal of Physics. Condensed Matter. 2015;**27**:313201. DOI: 10.1088/0953-8984/27/31/313201

Section 2

# Nanofluidics

Chapter 6

# Nanofluids and Computational Applications in Medicine and Biology

*Laith Jaafer Habeeb and Hasan Shakir Majdi*

## Abstract

The chapter comprises of two sections: the first concerns with the nanofluids, and the second is about the computational applications in medicine and biology. Nanotechnology is a novel logical methodology that includes materials and gear equipped for controlling the physical just as chemical properties of a substance at subatomic dimensions. This innovation can possibly expel the evident limits between biology, physics, and chemistry to some degree and shape up our present thoughts and comprehension. Consequently, numerous new difficulties and bearings may likewise emerge in education, research, and diagnostics in parallel by the extensive use of nanobiotechnology with the progression of time. Blood flow modeling in various arteries is an important topic of CFD biomechanics. Regardless of these endeavors and advances, there are as yet confounded inquiries around, for example, the interaction between blood flow and various artery diseases.

**Keywords:** nanofluidics, computational fluid dynamics, fluid-structure interaction, multiphysics systems coupling, multiphase flow

## 1. Introduction

The use of solid particles, like particles having millimeter or micrometer size, as additives suspended into the base fluid has been well recognized for numerous years. Nevertheless, they have not been of attention for the practical uses owing to problems, like the sedimentation that leads to increase the pressure drop in the channel of flow. The recent progress in the technology of material has done it able to engender an innovative nanofluid via the suspension of the particles having nanometer size in the base fluids that can vary the fluid movement and some properties of the base fluid. The nanofluids are solid-liquid composite alloys comprising solid nanofibers or nanoparticles having sizes distinctively from 1 to 100 nm suspended in a fluid. Various base fluids are commonly used. These are water, organic liquids (e.g., ethylene, triethylene glycols, refrigerants, etc.), oils and lubricants, bio-fluids, and polymeric solutions. The nanoparticles utilized in nanofluids include chemically stable metals (gold, copper, aluminum), metal oxides (alumina, silica, zirconia, and titania), metal carbides (SiC), metal nitrides (AIN, SiN), various forms of carbon (diamond, graphite, carbon nanotubes, fullerene), and functionalized nanoparticles. It is not a simple liquid-solid blend; the highly significant criterion of nanofluid is the agglomeration of freely steady suspension for long periods without resulting in any chemical variations in the base fluid. That can be done via increasing the liquid viscosity, via preventing the particles from agglomeration, and via using particles having nanometer size. The particle settling can prevent or minimize the density between solids and liquids [1].

Classical science and engineering disciplines already provide a wide, well-established base of knowledge for the understanding of these phenomena of nanofluids. Examples of areas that deal intensively with nanoscale phenomena include tribology, surface sciences, and colloid sciences (**Figure 1**).

Another classical research field that previously dealt with nanofluidic phenomena is the surface science, which studies the phenomena occurring at the interface of two phases, such as solid-liquid interfaces, solid-gas interfaces, and liquid-gas interfaces (**Figure 2**).

Computational fluid dynamics (CFD) is the using of computer-based simulation to analyze the systems that involve fluid flow, heat transfer, and connected phenomena. A numerical model is initially built utilizing a set of mathematical equations describing the flow. Then, such equations are solved employing a computer program for obtaining the flow parameters within the domain of flow. The development and application of CFD have undergone a considerable growth, and as a result, it has become a powerful tool in the design and analysis of engineering and other processes [2]. The governing equations of the models are partial differential equations (PDEs). Because the digital computers can merely recognize and handle the numerical data, such equations cannot be solved straightforward. Thus, the partial differential equations must be converted into numerical equations including merely the numbers and no derivatives. Such operation of making a numerical analogue to the partial differential equations is named "numerical discretization." The discretization operation includes an error because the "numerical" terms are merely the approximations to the initial "partial differential" terms. Such error, nevertheless, can be much reduced to low and thus acceptable levels. The main

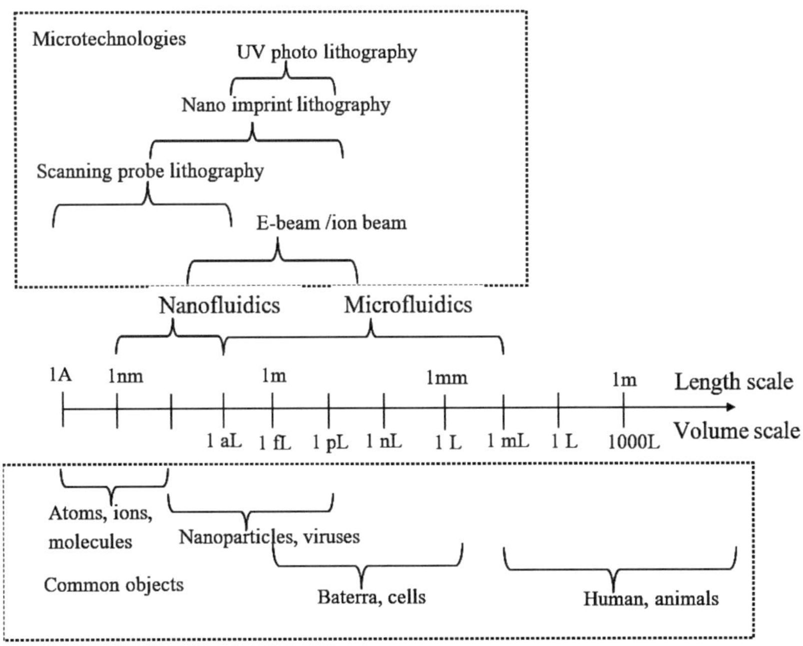


**Figure 1.**
*Length scales and volume scales of nanofluidics, microfluidics, common microtechnologies, and common objects [1].*

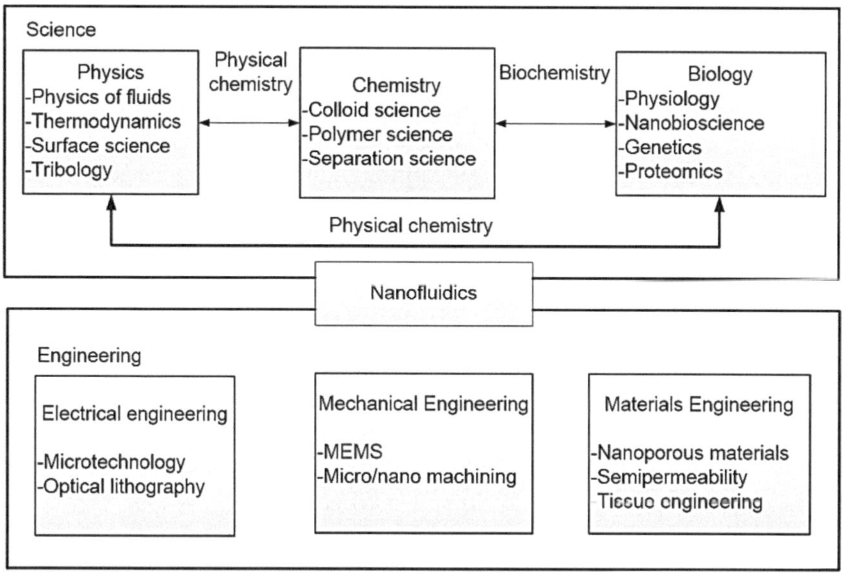


**Figure 2.**
*Classical areas of science and engineering related to nanofluidics [1].*

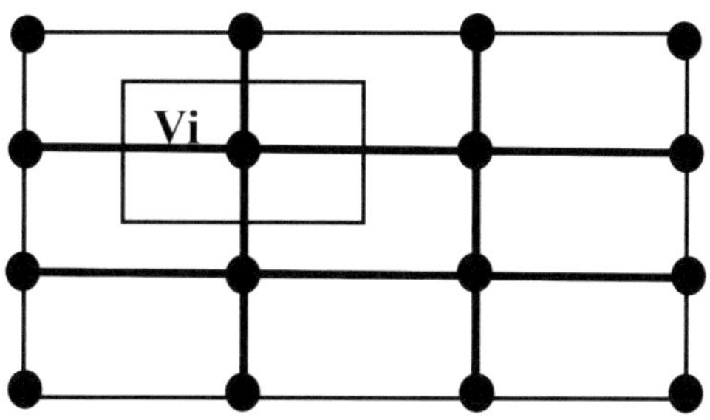


**Figure 3.**
*A control volume (Vi) surrounded by mesh elements.*

technique utilized for discretization is the "finite volume method." This method is likely the highly famous one employed for the "numerical discretization" in CFD. In some ways, it's similar to the "finite difference method (FEM)," but certain of its tools drag the features taken from the FEM. Such method includes discretization of spatial domain into the finite control volumes. The control volume overlaps with numerous mesh elements, and thus it can be split into sectors, each one backs to various mesh elements, as depicted in **Figure 3**.

The governing differential equations are integrated over every control volume. The obtained laws of integral conservation are precisely met for every control volume and for the whole domain, which is a discrete benefit of the FEM. Then, every integral term is changed into a separate form, therefore providing discretized equations at the nodal points, or centroids, of control volumes.

## 2. Some advantages and applications of using nanofluid and CFD

Due to the size of nanoparticles, the pressure drop is minimal, and a strong change in the properties of the main fluid, by the suspension of nanofluids and because of the size of nanoparticles, the liquid is considered one fluid.

According to the application, nanofluids are classified as heat transfer nanofluids, environmental (pollution cleaning) nanofluids, bio- and pharmaceutical nanofluids, and medical nanofluids (drug delivery, functional and tissue-cell interaction) [3]. The biomedical industry, for instant, the conventional method of cancer treatment kill the cells of cancers, drugs the radiation without damaging, cool the brain, and safe the surgery. Nanofluids can be utilized for cooling the equipment of welding and engines of automobiles and for cooling the high heat flux instrument, like a high-power laser diode array and high-power microwave tubes. Nanofluid can move throughout the tiny passage in MEMS to enhance efficiency. Within the industry of transportation, nanocars, General Motors (GM), are used. The nanofluid critical heat flux (CHF) measurement in a forced convection loop is beneficial for the nuclear uses. If nanofluid enhances the efficiency of the chiller by 1%, an electricity saving of 320 billion KWh or equivalent 5.5 million barrel of oil annually would be released in the USA only. Nanofluids possess the capability for the operations of deep drilling. Also, the nanofluid can be utilized to increase the dielectric power and age of an oil transformer via spreading nanodiamond particles [1].

There are numerous practical engineering problems for which one cannot determine the exact solutions. Such incapability to determine a perfect solution may be ascribed to either the intricate nature of the governing deferential equations or the difficulties that accrue from treating with the boundary and primary conditions. To treat with these problems, one resorts to numerical approximations. In contrast to the analytical solutions, which reveal the perfect behavior of a regime at any point through it, the numerical solutions approximate the perfect solutions merely at separate points. Two- and three-dimensional (CFD) modeling is time-consuming and computationally more expensive than one-dimensional analytical modeling. However, it can provide more information of the flow. It is also an effective alternative to experimental investigation. The simulation setup can be changed more flexibly than the experimental setup. Where is CFD used? It is used in aerospace, automotive, biomedical and chemical processing, HVAC (heating, ventilation, and air conditioning), hydraulics, marine, oil and gas, power generation, sports, etc. The major kinds of the fluid flow problems that the general-purpose CFD codes can solve are:

a. Kinds of flow: transient or steady, viscous or inviscid, laminar or turbulent (using a variety of turbulent models such as the k-model), compressible or incompressible, subsonic or supersonic speeds, or ultrasonic, multiphase (continuous phases or particles), chemical reacting, combustion, swirling, and non-Newtonian

b. Heat transfer modes: conduction, convection, and radiation

c. Kinds of material: solid (porous or homogenous) and fluid (gas or liquid)

d. Kinds of coordinate systems: cylindrical polar, Cartesian, curvilinear, moving/rotating, and body fitted

The use of computers is to help with all phases of engineering design work. Like computer-aided design (CAD), but also involving the construction and analysis of objects, the idea is to use computer processing and interactive computer graphics to enable engineers to create, modify, and analyze designs and hence to determine the structural, thermal, flow-field properties or other conditions of a regime. Computer-aided engineering (CAE) programs may employ a geometry definition from a CAD program as a starting point and always use some forms of FEA as the tool to conduct the analysis. The advantages of CAE system are:

a. It is capable of carrying out different engineering analyses, such as stresses and deformations, buckling, contact analyses, plastic deformations, vibration, heat transfer, fluid flow, magnetic field, coupled field problems, design optimization, etc.

b. It can work interactively with the CAD systems.

c. The analyses are facilitated through GUI (graphical user interface).

d. Different types of material properties can be included: isotropic, orthotropic, nonlinear, etc., and there is reduction of time.

e. Analysis and simulation can be modified and revised easily.

f. The results are presented graphically.

The disadvantages of CAE system are high cost of CAE software and special and advanced hardware, optical fatigue, and high cost of user's training and qualification.

## 3. Selected topics in nanofluidics and computational applications

Computational fluid dynamics studies are performed to grow a deeper insight into the field of the flow. In order to clarify the influence of the turbulence model, which involves the solution of a two-transport equation, model is used. Therefore, the techniques of the numerical solution will solve these Cartesian coordinate systems (x, y, and z). A three-dimensional geometry will generate.

### 3.1 Preparation of nanofluid and calculation of the thermal conductivity

Nanoparticles are made in one of two ways: physical processes and chemical processes. The physical techniques include mechanical grinding and the inert gas condensation technique. The chemical processes include chemical precipitation, spray pyrolysis, and thermal spaying. There are two ways to prepare nanofluid [1].

#### *3.1.1 Single-step method*

Single-step technique combines the production of nanoparticles and dispersion of nanoparticles into base fluid in a single step by the aid of chemical solvents [4].

#### *3.1.2 Two-step method*

This is the most widely used method for preparing nanofluids. This gives a large-scale production of nanofluids, whereas the single-step method is limited, in which dry powders are dispersed into a fluid. The second step is processing with the help of intensive magnetic force agitation, high-shear mixing, ultrasonic agitation, ball milling and homogenizing [4].

The main drawback in the two-step method is large agglomerations, whereas single-step method has limited agglomerations. The single-step method has the advantages in terms of controlling the particle size, reducing the particles agglomeration, and producing nanofluids containing metallic nanoparticles. The disadvantage is that it is difficult to prepare nanofluids with a high particle volume concentration [4].

The volume concentration is evaluated from the following relation in percentage:

$$\varphi = \frac{\text{volume of nanopartical}}{\text{volume of nanopartical} + \text{volume of water}} \times 100 \tag{1}$$

or

$$\varphi = \frac{\left(m/\rho\right)_{nanopartical}}{\left(m/\rho\right)_{nanopartical} + \left(m/\rho\right)_{water}} \times 100 \tag{2}$$

Due to difficulties of thermal conductivity measurement, it is estimated by the following equation:

$$K_{nf} = K_f \left[ \frac{2 + K_{pf} + 2\varnothing\left(K_{pf} - 1\right)}{2 + K_{pf} - \varnothing\left(K_{pf} - 1\right)} \right] \tag{3}$$

where

$$K_{pf} = \frac{K_p}{K_f} \tag{4}$$

In addition, thermal conductivity measurement techniques for nanofluids are transient hot-wire technique, transient plane source, thermal constant analyzer technique, thermal comparator, steady-state parallel-plate method, cylindrical cell method, temperature oscillation technique, 3ω method, and laser flash method.

## 3.2 The effect of nanofluid on heat transfer in a horizontal pipe

The enhancement of the distilled water heat transfer characteristics and the metal oxide nanofluid (ferrofluid)-type ($Fe_3O_4$) nanoparticles of average diameter

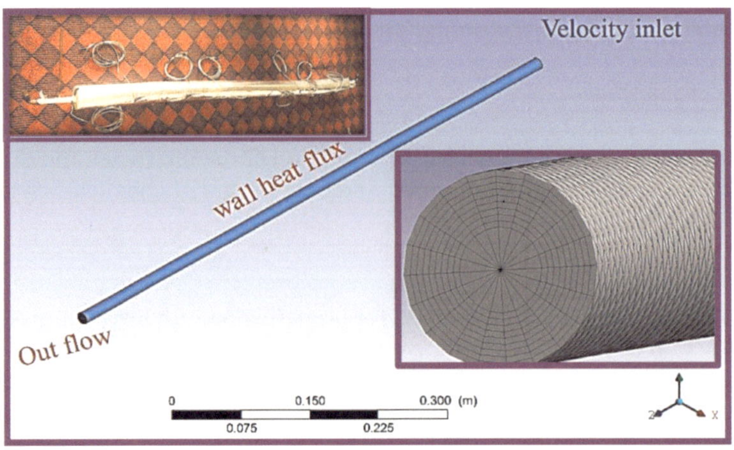


**Figure 4.**
*Geometry shape (experimental and numerical) and mesh generated.*

(80 nm) with distilled water at concentrations (φ) of 0.3, 0.6, and 0.9% by volume in a horizontal pipe have been studied experimentally and numerically [1]. All tests are conducted with the Reynolds number range of 2900–9820 and uniform heat flux 11,262–19,562 W/m$^2$. The numerical treatment of the present problem is based on the finite volume technique using commercial CFD software. The system geometry shown in **Figure 4** consists of a copper tube with a diameter of 1.4 cm and a length of 150 cm length. The fluid flows in the tube and is subjected to a uniform heat flux. The number of mesh element in this study is 305,492.

**Figure 5** shows the comparison between numerical and experimental results for water and ferrofluid with volume concentration (0 (water), 0.3, 0.6, and 0.9%). An agreement between the results was noticed, and the maximum division was 25, 29, 19, and 7% for nanofluid concentrations of 0, 0.3, 0.6, and 0.9%, respectively. This division could be related to the losses associated with the experimental part which are not taken into account theoretically, and one deals with it as a single-phase flow. However, both results have the same behavior. **Figure 6** shows the contours of temperature at the positions Z = 0, 0.22, 0.44, 0.66, 0.88, 1.1, 1.32, and 0.15 m for a volume concentration of 0.6% and Re = 5890. Such contours of temperature manifest increase in temperatures with decreasing ferrofluid concentration or with decreasing velocity.

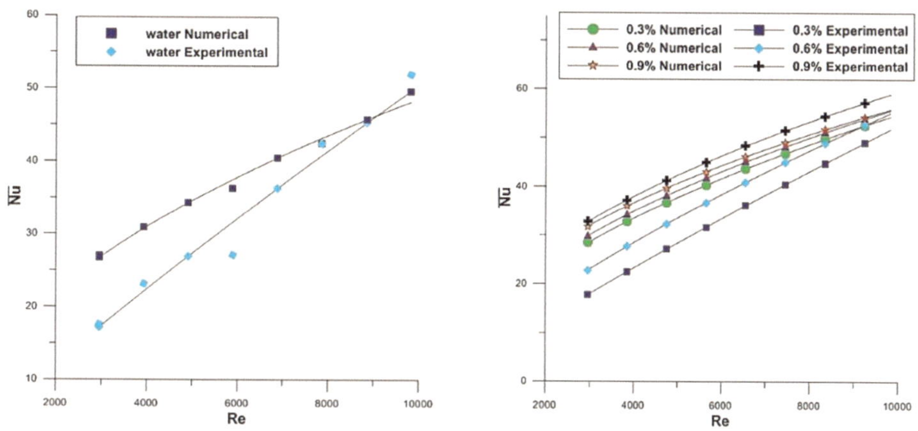


**Figure 5.**
*Comparison of numerical and experimental results for distilled water and ferrofluid.*

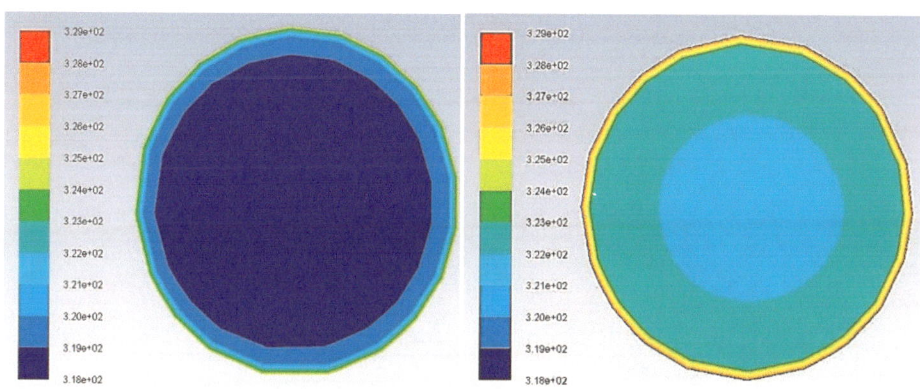


**Figure 6.**
*Temperature contours in K at locations of Z = 0.22 m (left) and 1.32 m (right) along the test section for ferrofluid of volume concentration = 0.6% with Re =5890.*

### 3.3 One- and two-way interaction study of nanofluid characteristics in a finned tube with twisted tape

This section presents an experimental and numerical study to investigate the improvement of the heat transfer and the interaction in a circular finned tube by utilizing one metal oxide [$\gamma$-$Al_2O_3$ (20 nm)]/distilled water nanofluid as a coolant with a typical twisted tape having a twist ratio (TR) of 1.85 [5]. The studied concentrations of nanofluids are $\varphi$ = 0, 3, and 5% by volume under laminar and turbulent flow conditions. The study includes constructing a test section that consists of aluminum tube of 1.5 m long, with internal and external diameters of 22 and 32 mm, respectively; see **Figure** 7. The coolant flows through the inner pipe under laminar

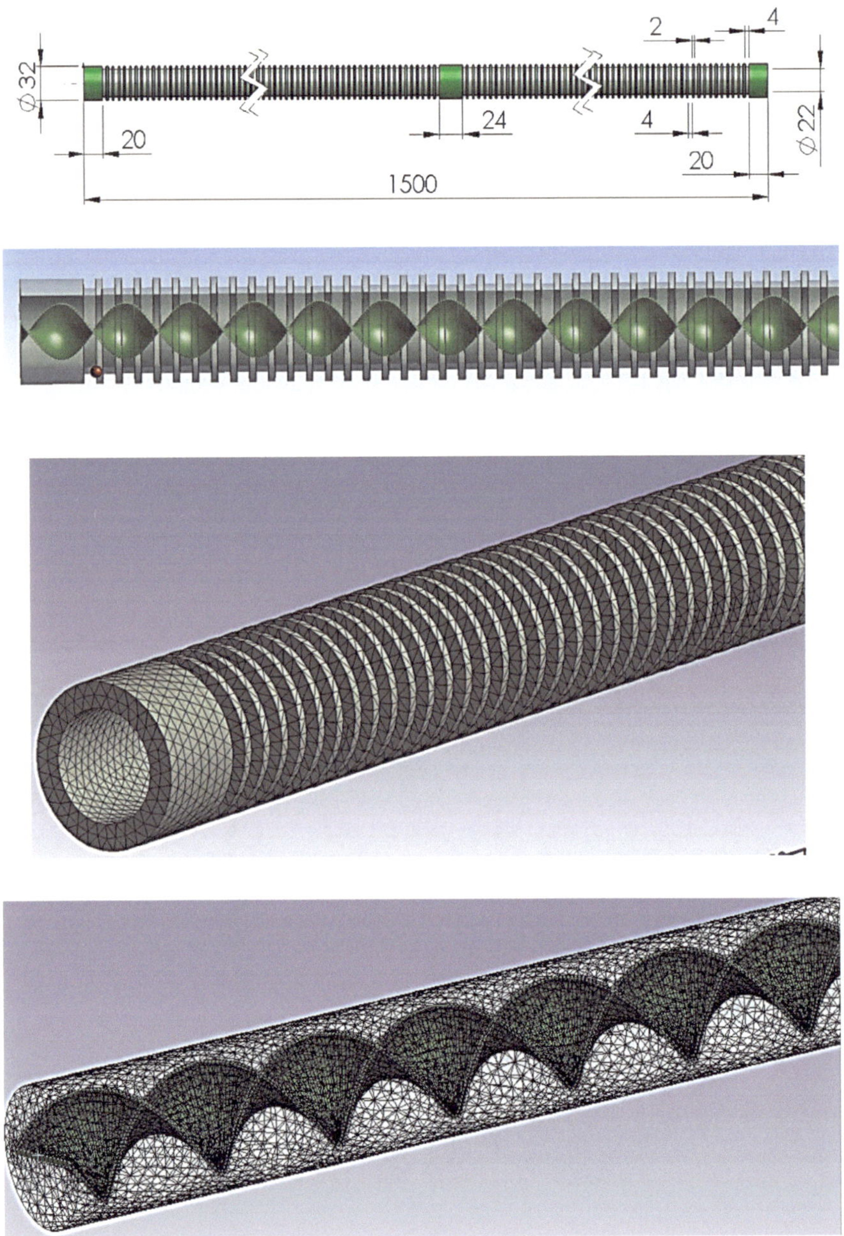


**Figure 7.**
*Physical geometry of finned tube (all dimensions in mm), geometry of twisted tape inside a finned tube insert, and meshing geometry.*

flow ($678 \geq Re \geq 2033$) and turbulent flow ($3390 \geq Re \geq 10,172$) regime with a constant inlet temperature of 60°C. Because of the complexity of twisted tape configurations and the one- and two-way fluid-structure interaction (FSI), it is impossible to determine an analytical solution of the governing equations for the practical configuration. The numerical simulations permit the intricate geometry analysis of the domain flow and the interaction by multiphysics systems coupling. Therefore, the commercial software of the finite volume numerical methods have been used to solve those equations and to study the interaction pattern of the fluid-heat-structure among fluid flow, typical twisted tape insert, and the finned tube having multidegrees of freedom flow-induced vibration for free and forced vibration.

The mathematical equations utilized for describing the fluid flow are continuity and momentum equations that characterize the conservation of mass and momentum. In addition, the momentum equations are recognized as the Navier-Stokes equations. For flows including heat transfer, another group of equations is needed for describing the energy conservation. Continuity equation is derived via employing the mass conservation principle to a small differential fluid volume. In the Cartesian coordinates, three equations with the following forms are determined. For laminar flow, the continuity, momentum, and energy equations are:

$$\frac{\partial u}{\partial x}+\frac{\partial v}{\partial y}+\frac{\partial w}{\partial z}=0 \tag{5}$$

$$\rho_{nf}\left(u\frac{\partial u}{\partial x}+v\frac{\partial u}{\partial y}+w\frac{\partial u}{\partial z}\right)=-\frac{\partial p}{\partial x}+\mu_{nf}\left(\frac{\partial^2 u}{\partial x^2}+\frac{\partial^2 u}{\partial y^2}+\frac{\partial^2 u}{\partial z^2}\right) \tag{6}$$

$$\rho_{nf}\left(u\frac{\partial v}{\partial x}+v\frac{\partial v}{\partial y}+w\frac{\partial v}{\partial z}\right)=-\frac{\partial p}{\partial y}+\mu_{nf}\left(\frac{\partial^2 v}{\partial x^2}+\frac{\partial^2 v}{\partial y^2}+\frac{\partial^2 v}{\partial z^2}\right) \tag{7}$$

$$\rho_{nf}\left(u\frac{\partial w}{\partial x}+v\frac{\partial w}{\partial y}+w\frac{\partial w}{\partial z}\right)=-\frac{\partial p}{\partial z}+\mu_{nf}\left(\frac{\partial^2 w}{\partial x^2}+\frac{\partial^2 w}{\partial y^2}+\frac{\partial^2 w}{\partial z^2}\right) \tag{8}$$

$$\left(\rho C_p\right)_{nf}\left(u\frac{\partial T_{nf}}{\partial x}+v\frac{\partial T_{nf}}{\partial y}+w\frac{\partial T_{nf}}{\partial z}\right)=K_{nf}\left(\frac{\partial^2 T_{nf}}{\partial x^2}+\frac{\partial^2 T_{nf}}{\partial y^2}+\frac{\partial^2 T_{nf}}{\partial z^2}\right) \tag{9}$$

For turbulent flow, the continuity, momentum, and energy equations and turbulence kinetic energy ($k$) and its dissipation rate ($\varepsilon$) are:

$$\frac{\partial \overline{u}}{\partial x}+\frac{\partial \overline{v}}{\partial y}+\frac{\partial \overline{w}}{\partial z}=0 \tag{10}$$

$$\begin{aligned}&\left(\overline{u}\,\frac{\partial\overline{u}}{\partial x}+\overline{v}\,\frac{\partial\overline{u}}{\partial y}+\overline{w}\,\frac{\partial\overline{u}}{\partial z}\right)+\left(\frac{\partial}{\partial x}\left(\overline{u'^2}\right)+\frac{\partial}{\partial y}\left(\overline{u'v'}\right)+\frac{\partial}{\partial z}\left(\overline{u'w'}\right)\right)\\&=-\frac{1}{\rho_{nf}}\frac{\partial p}{\partial x}+\gamma\,\nabla^2\overline{u}\end{aligned} \tag{11}$$

$$\begin{aligned}&\left(\overline{u}\,\frac{\partial\overline{v}}{\partial x}+\overline{v}\,\frac{\partial\overline{v}}{\partial y}+\overline{w}\,\frac{\partial\overline{v}}{\partial z}\right)+\left(\frac{\partial}{\partial x}\left(\overline{u'v'}\right)+\frac{\partial}{\partial y}\left(\overline{v'^2}\right)+\frac{\partial}{\partial z}\left(\overline{v}\overline{w'}\right)\right)\\&=-\frac{1}{\rho_{nf}}\frac{\partial p}{\partial y}+\gamma\,\nabla^2\overline{v}\end{aligned} \tag{12}$$

$$\begin{aligned}&\left(\overline{u}\,\frac{\partial\overline{w}}{\partial x}+\overline{v}\,\frac{\partial\overline{w}}{\partial y}+\overline{w}\,\frac{\partial\overline{w}}{\partial z}\right)+\left(\frac{\partial}{\partial x}\left(\overline{u'w'}\right)+\frac{\partial}{\partial y}\left(\overline{v'w'}\right)+\frac{\partial}{\partial z}\left(\overline{w'^2}\right)\right)\\&=-\frac{1}{\rho_{nf}}\frac{\partial p}{\partial z}+\gamma\,\nabla^2\overline{w}\end{aligned} \tag{13}$$

$$\left(\overline{u}\,\frac{\partial\overline{T}_{nf}}{\partial x}+\overline{v}\,\frac{\partial\overline{T}_{nf}}{\partial y}+\overline{w}\,\frac{\partial\overline{T}_{nf}}{\partial z}\right)=\alpha\,\nabla^2\overline{T}_{nf}+\left(-\frac{\partial}{\partial x}\left(\overline{u'T'_{nf}}\right)-\frac{\partial}{\partial y}\left(\overline{v'T'_{nf}}\right)-\frac{\partial}{\partial z}\left(\overline{w'T'_{nf}}\right)\right) \tag{14}$$

$$\frac{\partial}{\partial t}\left(\rho_{nf}k\right)+\frac{\partial}{\partial x_i}\left(\rho_{nf}ku_i\right)=\frac{\partial}{\partial x_j}\left[\left(\mu+\frac{\mu_t}{\sigma_k}\right)\frac{\partial k}{\partial x_j}\right]+G_k+G_b-\rho_{nf}\varepsilon-Y_M+S_k \tag{15}$$

$$\frac{\partial}{\partial t}\left(\rho_{nf}\varepsilon\right)+\frac{\partial}{\partial x_i}\left(\rho_{nf}\varepsilon u_i\right)=\frac{\partial}{\partial x_j}\left[\left(\mu+\frac{\mu_t}{\sigma_k}\right)\frac{\partial \varepsilon}{\partial x_j}\right]+C_{1\varepsilon}\frac{\varepsilon}{k}\left(G_k+C_{3\varepsilon}G_b\right)-C_{2\varepsilon}\frac{\varepsilon^2}{k}+S_\varepsilon \tag{16}$$

Grid independence test is carried out to obtain the most suitable computational grid for which a finer grid provides the similar outputs with the initial one, and the outputs do not vary as grid becomes finer. The checking procedure, either the solution is grid independent or not, is to generate a grid with many cells for comparing the solutions of both models. The tests of grid refinement for Nusselt number explain that the grid size of almost 2 million cells provides a sufficient accuracy and resolution to be adopted as the standard for all cases. The grid independence test performed for typical twist tape with TR = 1.85 configuration is shown in **Figure 8**.

The analysis of fluid-structure interaction (FSI) is an instant of a multiphysics problem, where the interaction between two dissimilar analyses is taken into consideration. This analysis includes conducting a structural analysis taking into consideration the interaction with the corresponding fluid analysis. The interaction between both analyses distinctively occurs at the model solution boundary

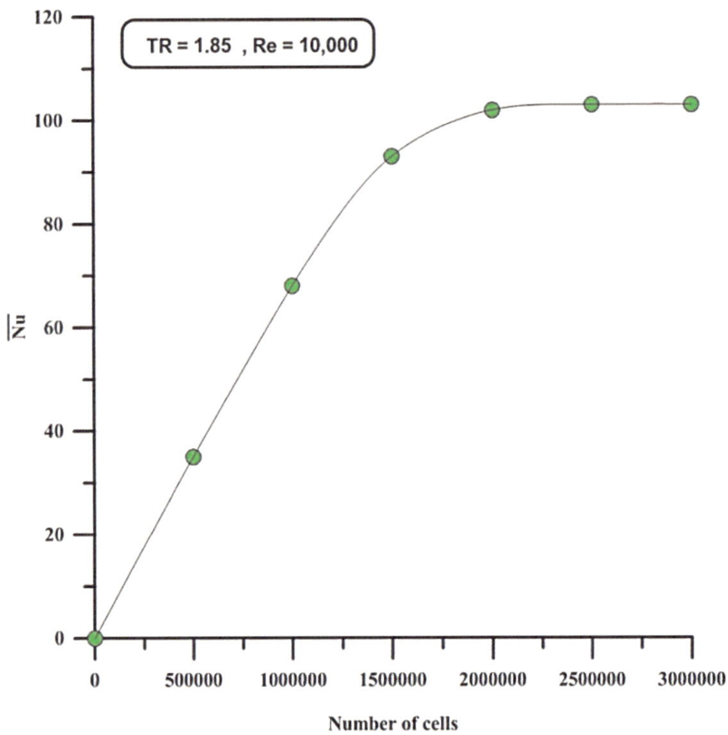


**Figure 8.**
*The grid-independent solution test for typical twisted tape.*

(the fluid-structure interface), where the outputs of one analysis are passed to the other one as a load. There are two different fluid-structure interaction approaches that can be used, depending on the physical nature of the interaction. The multiphysics problems are too hard to solve via the analytical approaches. Accordingly, they must be solved either via employing experiments or numerical simulations. The progressed methods and the existence of the reckoned commercial software in both CFD and computational structural mechanics (CSM) have made such numerical simulation possible. There are two dissimilar methods to solve the problems of FSI utilizing such software: the monolithic method and the partitioned method. In one-way coupled FSI, the results (forces) from the fluid analysis at the fluid-structure interface are applied as a load to the structural analysis. The boundary displacement from the structure is not passed back to the fluid analysis. The assumption is that the deformation of the structure is small, having insignificant effect on the fluid flow prediction. This allows the fluid analysis and structure analysis to be run independently. This technique will be used for the theoretical analysis. In the two-way coupled FSI, the structural analysis results are conveyed to the fluid analysis as a load. In a similar way, the fluid analysis results are passed back to the structural analysis as a load. For instance, the fluid pressure at the boundary can be applied as a load on the structural analysis, and the resulted displacement, velocity, or acceleration determined in the structural simulation could be passed on as a load to the fluid analysis. The analysis will carry on till the whole equilibrium (convergence) is attained between the fluid flow solution and the structural solution.

The simulated values of average Nusselt number are compared with the experimental results, as shown in **Figure 9**. The computed values agree with the experimental data within ±13 and ±12% for laminar and turbulent flow, respectively. The simulated friction factors along the tube against the Reynolds no. are also compared to those determined from the experiments. It's noticed that the simulated data are matching with the experimental data within ±11% for the average Nusselt no. and ±9% for the friction factor, respectively. However, both results have the same behavior, and the differences are with acceptable values. **Figure 10** shows the velocity vector at location of Z = 0.5 m along the test section. A secondary flow is created, and a rotational movement is noted along the tube, and this will enhance the heat transfer within the tube.

**Figure 11** elucidates the 3D view for the distribution of static temperature along the test section at the midplane (environment and tube) for a nanofluid having a 3%

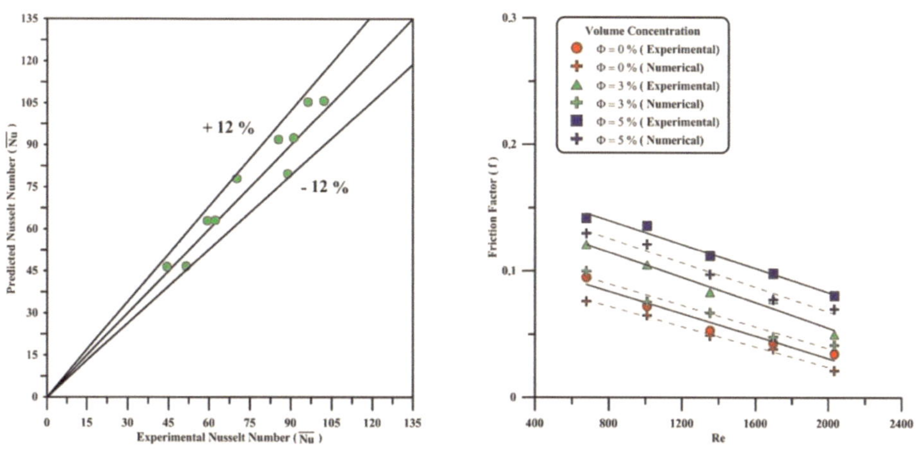


**Figure 9.**
*Comparison of experimental and predicted* $\overline{\mathbf{Nu}}$ *for turbulent flow and for Re and f for laminar flow.*

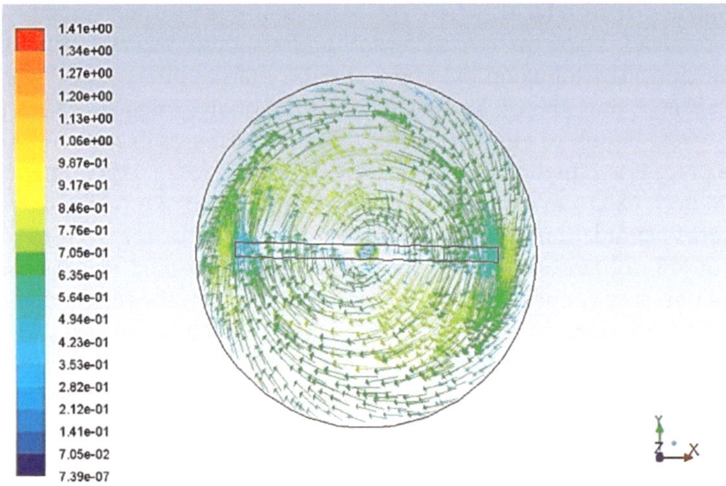


**Figure 10.**
*Velocity vectors in m/s at Re =10,172 for nanofluid (φ = 3%) at Z = 0.5 m.*

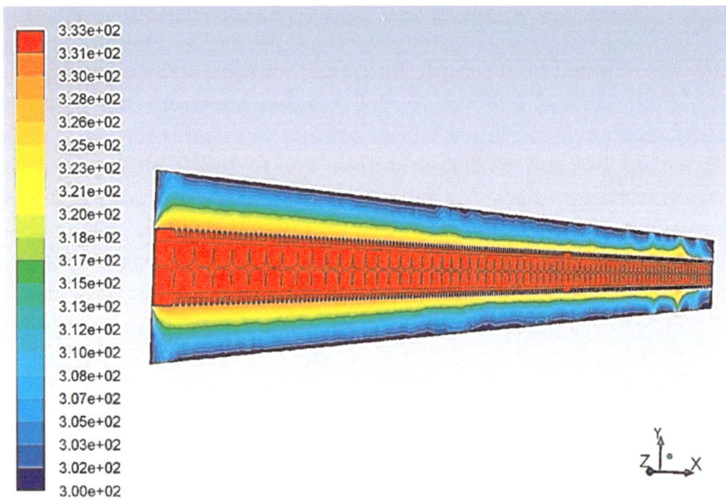


**Figure 11.**
*Temperature distribution along the test section for φ = 3%.*

volume concentration and a Reynolds no. equal to 5086 for a turbulent flow. **Figure 12** shows the test section deformation calculated using static structural-mechanical is a solution processing model, under the influence of enlarging its value ($1.8 \times 10^3$ autoscale). The maximum deformation occurs at the beginning of the tube in all models because of the high temperature concentration at this region. From this figure, one can see that the total deformation decreases as the volume concentration and the mass flow rate increase due to the frequency effect, which is decreased with the increase in nanoparticle mass.

**Figure 13** highlights the 3D view for the twisted tape with the velocity vector along a focused distance of the test section for a nanofluid having a 3% volume concentration. In this figure, one can see that the vector magnitude and direction change at different periods of time due to effect of the twisted tape deformation, which is much higher than that of the tube, resulted from thermal expansion (elongation), fluid pressure effect, and reaction force on it.

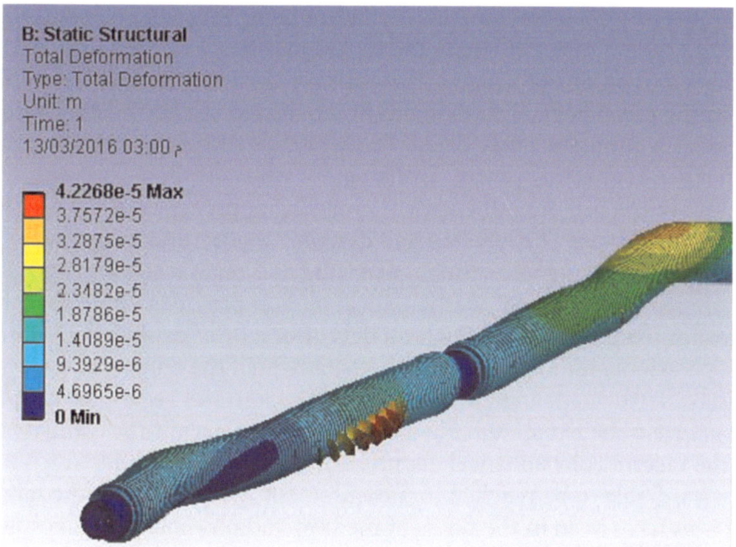


**Figure 12.**
*Total deformation from one-way interaction for Re = 10,172 and φ = 5%.*

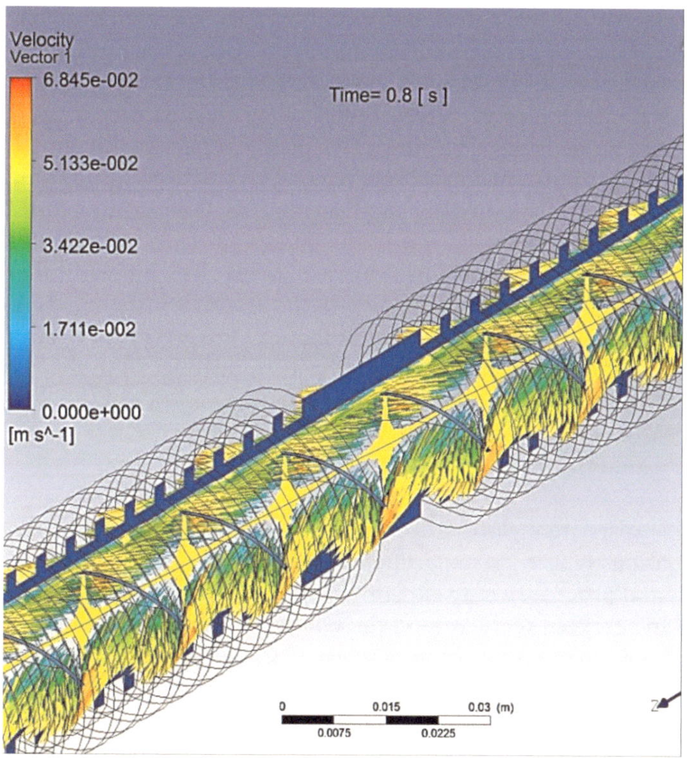


**Figure 13.**
*Velocity vectors for two periods at Re = 678 and φ = 3% and T2 = 0.8 s.*

## 4. Selected topics in medicine and biology

During the previous decades, there has been a significant raise in the use of quantitative techniques for studying the physiological regimes. Recent methods for conducting the physiological measurements are being steadily evolved and used, and there has been a relevant rise in the techniques that exist for analyzing and interpreting the data of experiment. Increasingly, such methods are obtaining their way within the physiological studies and in the related studies in clinical sciences and medicine. A supplementary driver for the whole of this is, certainly, the existence of further computing power. Utilizing the whole of these gathered is causing an increment in the use of mathematical modeling approaches in the physiological studies. The further use of modeling and dynamic regime analysis provides advantages for the biomedical engineering, governing and regime science, and physiology. The proper application of mathematical models provides numerous potential advantages for the physiologist. Such models offer a brief explanation of intricate dynamic operations, indicating the methods, in which the enhanced experimental design can be performed, and empowering the hypotheses regarding the physiological structure to be examined. Further to that, they permit the estimates to be done for the factors (physiological quantities) that are in different way not straightforward able to be reached to measurement. Despite firstly the most modeling uses have been in the fields of medical and physiological investigation, they are presently further being utilized as assistances in diagnosing and treating the disease [6]. The biomedical engineering (BME) is an engineering branch involved with solving problems in the biology and the medicine. Biomedical engineers use principles, methods, and approaches drawn from the more traditional branches of electrical, mechanical, chemical, material, and computer engineering to solve this wide range of problems. They use them with other fundamentals to the problems in the fields of life sciences and healthcare, i.e., this engineer must also be familiarized with the biological ideas of physiology and anatomy at the cellular, molecular, and regime levels. Practicing the healthcare needs the familiarization with the nervous system, cardiovascular regime, circulation, respiration, body fluids, and kidneys. The biomedical engineering field is expanding fast. The biomedical engineers will take a big role in the investigation in the life sciences and device evolution for the adequate healthcare delivery. The biomedical engineering scope ranges from the bionanotechnology to the assisting instruments, from the cellular and molecular engineering to the robotics of surgery, and from the neuromuscular regimes to the synthetic lungs. The ideas introduced in this context will assist the biomedical engineers to operate in such variant field [7].

### 4.1 Dentistry analysis

Dental scientists are making increased usage of computational methods, particularly in situations where the experimental procedures fail to give proper answers. An experimental procedure may explain the maximum load of a tooth failure, but it cannot give an accepted reply around the failure evolution mechanism. Dentistry analysis is done in many ways, such as stress analysis, fluid mechanics and dynamic analysis, thermal analysis, restorative material analysis, and so on. The structure of the normal tooth conveys the loads of the external biting via the enamel within the dentin. Since the teeth aren't stiff structures, so they subject to deformation (strain) during the usual loading. The focused external loads are spread over a big internal volume of the tooth structure, and thus the local stresses are less. Within such operation, a little quantity of the dentin deformation may take place that causes the tooth bending. If a load is exerted, the structure is subject to a deformation since its bonds are sheared, stretched, or compressed. As the loading progresses, this structure will deform. Firstly, such deformation (strain) is totally a reversible elastic strain. However, the incremented loading eventually makes also certain irreversible strain (plastic strain) that results in a fixed deformation.

The onset of the plastic strain point is named the elastic limit (proportional limit, or yield point). This point is exhibited on the stress-strain curve at the point, where the straight line begins to be curved. Thus, progress of the plastic strain ultimately results in a failure via fracture. The largest stress prior to fracture is the maximum strength, and the whole tensile strain (plastic) at the fracture is named the elongation. **Figure 14** shows the sound and restored teeth models with finite element mesh. Since the enamel is of greater stiffness than that of the dentin, it will take most of the applied load and distributes it all over the dentin in a uniform manner. In this case, only small values of stress will reach the dentin. Whereas, in **Figure 15A**, Young's modulus values of the enamel are assumed to be high, the load is applied at the tip of the buccal cusp. The enamel is acting here as a stress distributor, where the stress would transfer in a shape very similar to the stressed enamel. Moreover, when Young's modulus value of the enamel is low, the stress tunnels through the enamel in a sharp manner, reaching the dentin, which is assumed to be of higher Young's modulus value (**Figure 15B**). Then, the dentin in turn would act as a stress distributor when transmitting it to the following parts and the pulp [8].

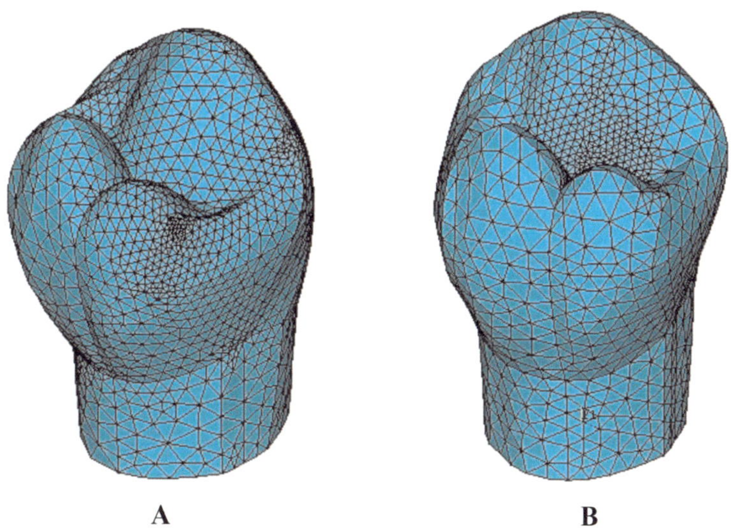


**Figure 14.**
*Mesh of (A) sound model and (B) restored model.*

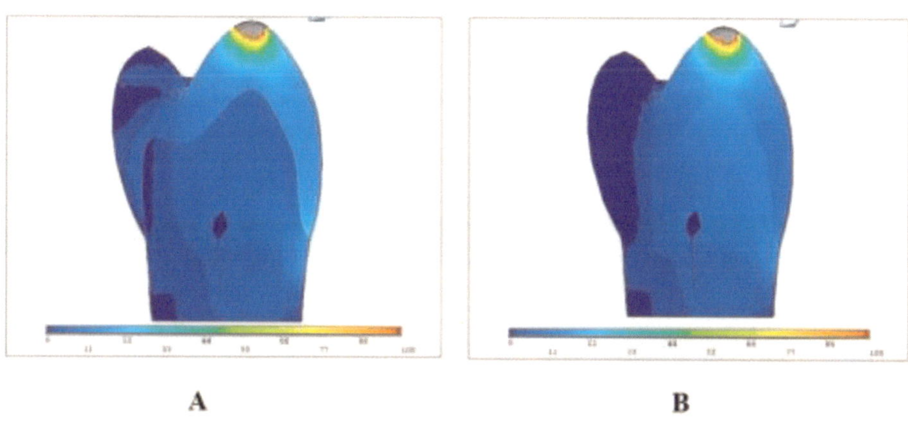


**Figure 15.**
*Distribution of the von Mises stress contour of the sound tooth model subjected to loading case.*

## 4.2 Contaminant degradation

One of the specific bioremediation mechanisms is the contaminant degradation in the soil via the plant enzymes that are exuded from the roots. For the soil that is contaminated by petroleum, the result of bioremediation is suggested to be depended upon the degrading microorganism stimulation in the rhizosphere, named rhizodegradation or phytostimulation. The biodegradation is commonly a slow operation owing to the contaminant's hydrophobic nature and the resulted bioavailability limitations. The petroleum hydrocarbons, like diesel with the n-alkane markers that range in size from C8 to C25, are mostly decreased organic molecules that can work as a carbon origin and electron donor for the microorganisms, for supporting the microbial metabolism. The hydrocarbon biodegradation reduces with the raise of the molecular weight. Microorganisms are able to degrade the hydrocarbons with a broad range of n-alkanes between C10 and C35, among which C14–C19 are desired. Beneath the anaerobic circumstances, the electron acceptors other than $O_2$ are utilized for the microbial respiration and through such operation; hydrocarbons are oxidized to the intermediate molecules and finally to $CO_2$, whereas the terminal electron acceptors are decreased. Rhizobacteria (RB) are characterized as the bacteria that live in the surrounding area of the root or on the surface of the root. The hydrocarbon degradation is enhanced via a rhizosphere influence with plants that exude the organic constituents throughout their roots, affecting the variety, abundance, or the ability of potential hydrocarbon to degrade the microorganisms in the region that surrounds the roots. The roots provide suitable attachment locations for the microorganisms and also provide the nutrients in the shape of exudates composing of organic acids and amino acids, sugars, enzymes, and intricate carbohydrates. Moreover, the root exudates from plants do help to degrade the toxic organic chemicals and acts as substrates for the soil microorganisms to increase the biodegradation rate of the organic contaminants. The hydrocarbon-contaminated soil biodegradation that exploits the capability of microorganisms for degrading and/or detoxifying the organic contamination has been built as an adequate, versatile, economic, and environmentally a good processing for the kerosene-contaminated soils. The microorganisms make biosurfactant being plentiful in nature; they hinder the water (groundwater, seawater, and freshwater) and the land (sediment, sludge, and soil). Additionally, they can be obtained in the utmost surroundings (e.g., reservoirs of oil) and prosper at a broad range of salinity, temperatures, and pH values. Nevertheless, the microbial communities of hydrocarbon-degrading abide the highly proper ambient for a broad capability for the production of biosurfactant. The hydrocarbon-degrading bacterial populations are, in general, prevailed via a few major genera, including *Sphingomonas*, *Bacillus*, *Actinobacteria* in sediments and soils, *Pseudomonas* and *Klebsiella*, and *Halomonas*, *Alcanivorax*, *Acinetobacter*, and *Pseudoalteromonas* in the marine ecosystems. It has been documented that 2–3% of the screened populations within the uncontaminated soils are microorganisms that produce biosurfactant. That raises to 25% in the polluted soils. From the other side, the methods of enrichment culture, specifically for the hydrocarbon-degrading bacteria, may result a greater detection of the biosurfactant makers with estimates till 80%. The biosurfactants made via microorganisms are divided into two various classes depending upon their chemical composition: like the surface-active agents with less molecular weight named biosurfactants and the biosurfactants with more molecular weight denoted as bioemulsifiers.

## 4.3 Wastewater treatment

One of the multiphase flow applications is the three-phase fluidized bed (gas-liquid-solid fluidized bed) which has appeared recently as one of the major

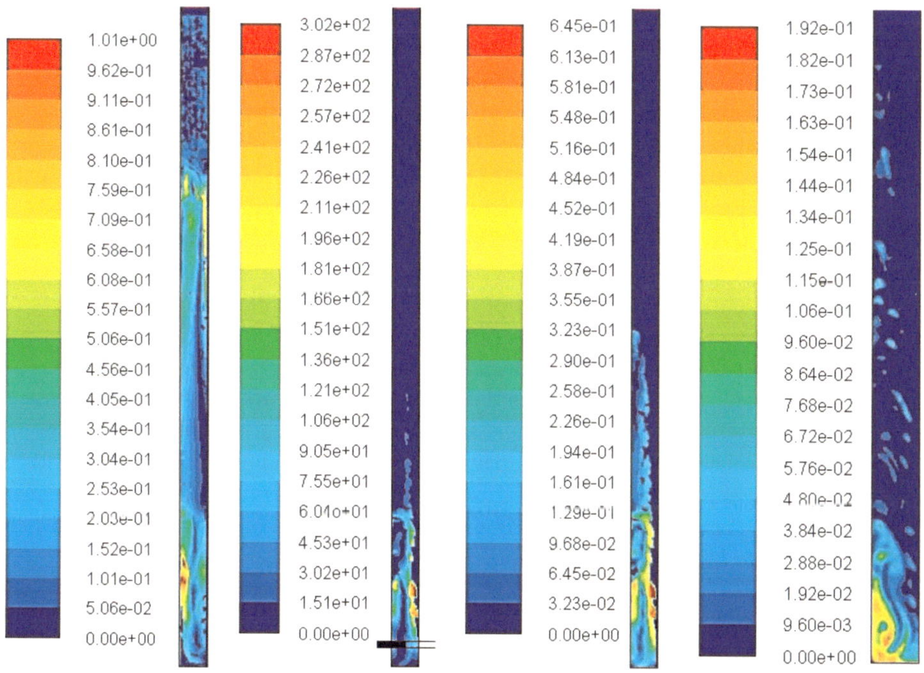


**Figure 16.**
*Contours of velocity magnitude for air in m/s, contours of dynamic pressure for solid particles in Pascal, contours of velocity magnitude for solid particles in m/s, and contours of volume fraction for solid particles, respectively.*

promising instruments for the three-phase process. This instrument is of important industrial significance as proofed from its broad use in the chemical, biochemical, and petrochemical treatment. The fluidized beds work in numerous aims in the industry, like promoting the catalytic and non-catalytic reactions. Three-phase fluidized beds have been used adequately in numerous industrial operations, like in the $H_2$-oil operation for the residual oil hydrogenation and hydrodesulfurization; H-coal operation for the coal liquefaction; Fischer-Tropsch operation; bio-oxidation process for wastewater treatment; biotechnological operations, such as pharmaceuticals and mineral industries; fermentation and aerobic wastewater processing; and so on. One of the recent biotechnological process applications is the study of three-phase fluidized beds for dried algae such as chlorella after they are mixed, crushed, dried, and immobilized to us as the solid phase. The liquid phase is the water, and the gas phase is the air. **Figure 16** represents (from left to right) the contours of velocity magnitude for air in m/s at time = 3 s, contours of dynamic pressure for solid particles in Pascal at time = 3 s, contours of velocity magnitude for solid particles in m/s at time = 3 s, and contours of volume fraction for solid particles at time = 3 s, respectively.

## 5. Conclusion

Nanofluids, as mentioned earlier, are prepared from suspending nanoparticles into dilute liquid. The thermal behavior of nanofluids may offer a huge invention for heat transfer. Too many applications are in field of nanofluidics: transportation, electronics cooling, nuclear systems cooling, boiler flue gas temperature reduction, energy efficient cooling, heating of buildings without increased pumping power in heating, ventilation and air conditioning, heat exchangers, biomedical industry, for example, traditional cancer treatment method, kill cancers cells, drugs radiation without damaging, cool the brain, safer surgery, heat pipes, fuel cell, solar water heating, domestic refrigerator, diesel combustion, thermal storage, etc. Solving CFD problem usually consists of four components: geometry and grid generation, setting up a physical model, solving it, and post-processing the computer data. The created geometry and grid are generated, the set problem is computed, and the way acquired data is presented is very well known. Precise theory is available. Mathematical modeling is now widely applied in physiology and medicine to support the life scientist and clinical worker. Mathematical modeling finds application in medical research, in education, and in supporting clinical practice. The use of models can, for example, yield quantitative insights into the manner in which physiological systems are controlled. In the educational setting, medical students can use computer model simulation to explore the dynamic effects of pathophysiological processes or of drug therapy. In the clinical arena, mathematical models can enable estimates to be made of physiological parameters that are not directly measurable—useful for example in diagnosis, as well as enabling predictions to be made as to how changes in drug therapy will impact on variables of clinical importance such as blood pressure or blood glucose concentration.

## Acknowledgements

The authors would like to thank the University of Technology and Al-Mustaqbal University College for the support in the present work.

## Nomenclature

| | |
|---|---|
| $Cp_{nf}$ | specific heat of nanofluid at constant pressure (kJ/kg.K) |
| $k_{nf}$ | thermal conductivity of the nanofluid (W/m.K) |
| $p$ | pressure ($N/m^2$) |
| $T$ | temperature (°C) |
| $u, v, w$ | velocity component in Cartesian coordinate (m/s) |
| $x, y, z$ | Cartesian coordinate (m) |

### Greek symbols

| | |
|---|---|
| $\varepsilon$ | dissipation rate (1/s) |
| $k$ | turbulent kinetic energy ($m^2/s^2$) |
| $\rho_{nf}$ | density of the nanofluid ($kg/m^3$) |
| $\varphi$ | volume fraction (Vol.%) |
| $\nabla$ | represents the partial derivative of a quantity with respect to all directions in the chosen coordinate system (—) |

### Subscripts

| | |
|---|---|
| $i, j, k$ | tenser indices |
| $nf$ | nanofluid |

### Superscripts

| | |
|---|---|
| $( )'$ | fluctuation component |

## Author details

Laith Jaafer Habeeb[1]* and Hasan Shakir Majdi[2]

1 Training and Workshops Center, University of Technology, Baghdad, Iraq

2 Al-Mustaqbal University College, Babylon, Hillah, Iraq

*Address all correspondence to: laithhabeeb1974@gmail.com

## References

[1] Asker AH. The effect of magnetic field with nanofluid on heat transfer in a horizontal pipe [thesis]. Baghdad-Iraq: Mechanical Engineering Department, University of Technology; 2016

[2] Versteeg HK, Malalasekera W. Introduction to Computational Fluid: The Finite Volume Method. England: Longman Scientific & Technical; 1996

[3] Cheng L. Nanofluid Heat Transfer Technologies. Recent Patents on Engineering. Bentham Science Publishers; Vol. 3, Number 1, 2009. pp. 1-7(7)

[4] Yu W, Xie H. A review on nanofluids: Preparation stability mechanisms, and applications. Journal of Nanomaterials. 2012;**2012**:1-17. Article ID 435873

[5] Magel BM. One and two-way interaction study of nanofluid characteristics in a finned tube with twisted tape [thesis]. Baghdad-Iraq: Mechanical Engineering Department, Faculty of Engineering, Al-Mustansiriayah University; 2016

[6] Cobelli C, Carson E. Introduction to Modeling in Physiology and Medicine. 1st ed. Amsterdam: Academic Press is an imprint of Elsevier; 2008

[7] Dunn SM, Constantinides A, Moghe PV. Numerical Methods in Biomedical Engineering. Amsterdam: Academic Press is an imprint of Elsevier; 2006

[8] Abdul-Khaliq Hussain W. Stresses analysis of sound and restored teeth [thesis]. Baghdad-Iraq: Department of Applied Sciences, University of Technology; 2007

Chapter 7

# Convection Currents in Nanofluids under Small Temperature Gradient

*Jyoti Sharma and Urvashi Gupta*

## Abstract

Nanobiotechnology has huge number of applications in medical science thereby improving health care practices. Keeping in mind the applications of nanoparticles and the convection patterns in biological fields, behaviour of nanofluids is explored for small temperature difference in the layer. The flow of nanofluids is usually described by system of differential equations. A mathematical model for the system based on conservation laws of mass, momentum and energy is formed. To get the insight of the problem, complex equations are simplified wherever needed to get interesting results without violating the necessary physics. The influence of physical properties such as density and conductivity of metallic/non-metallic nanoparticles is examined on the onset of convection currents in the fluid layer.

**Keywords:** nanofluids, natural convection, conservation equations, metallic and non-metallic nanoparticles

## 1. Introduction

In 1959, the celebrated physicist and Nobel laureate Richard Feynman presented an idea of nanotechnology in his talk "There is a plenty of room at the bottom-An invitation to enter a new field of physics" by emphasizing on the fact that the laws of physics allow us to arrange the atoms the way we want. Almost a century ago, Maxwell [1] initiated working on this issue theoretically and unveiled that the particles of size of micrometer and millimeter, if used in traditional fluids can resolve the motive in a more efficient manner. Yet they had few drawbacks like clogging, erosion in micro channel and settling down which were curbed with the evolution of better substitute; nanosized particles (called as nanoparticles). The suspension of nanoparticles in the regular fluids comprised the nanofluids [2]. Nanofluids have also shown many interesting properties, and the distinctive features (refer **Table 1**) resulting in unprecedented potential for many applications particularly in biological, medical and biomedical applications.

The catalytic role of nanoparticles in intensifying the thermal conductivity of nanofluids is analyzed by many researchers: Masuda et al. [3], Eastman et al. [4], Das et al. [5] and others. In 2006, Buongiorno [6] pioneered the formulation of conservation equations of nanofluids by incorporating the impacts of diffusion due to Brownian motion and thermophoresis of nanoparticles. He made an observation that the velocity of nanoparticles can be perceived as a sum of base fluid and

| Properties | Microparticles | Nanoparticles |
|---|---|---|
| Stability | Not stable | Stable |
| Surface/Volume ratio | One | One thousand times more than that of microparticles |
| Conductivity | Less | High |
| Clogging | More | Negligible |
| Erosion | Yes | No |
| Nanoscale phenomenon | No | Yes |

**Table 1.**
*Comparison of particles.*

relative (slip) velocities. To prosecute his research, he considered seven slip mechanisms; inactivity, magnus effect, Brownian motion, diffusiophoresis, thermophoresis, gravitational settling and fluid drainage. Throughout his investigation, he agreed that from all these seven techniques, Brownian diffusion and thermophoresis have a significant part in the absence of turbulent effects. Choi et al. [7] found that carbon nanotubes provide highest thermal conductivity enhancement of nanofluids. There are ample number of evaluations on thermal conductivity of nanofluids [8–11] in which they discussed and analyzed the theoretical as well as experimental results. Heat transfer in nanofluids because of convection has been examined and contemplated by Das and Choi [12], Ding et al. [13] and Das et al. [14].

The ballistic character of heat transfer within nanoparticles has been studied by Chen [15]. Abnormal increase in viscosity is generally observed in relation to the base fluid. The presence of nanoparticles has found to enhance thermal conductivity [4, 7, 16–19]. At very low nanoparticle volume fractions ($<0.1\%$), a heat transfer enhancement up to 40% has been reported [8] and this percentage is found to enhance with temperature [5] and concentration of nanoparticles [16]. The results of Choi et al. [7] established the unexpected non-linear character of measured thermal conductivity with nanotube loadings at low concentration while all theoretical studies concluded a linear relationship. Also, it was discovered that thermal conductivity strongly depends on temperature [5] and particle size [20]. Pak and Cho [21] in their study also reported the heat transfer data for turbulent flow of nanofluids having nanoparticles as aluminum and titanium in circular tubes. They found that Nusselt number is up to 30% more than that of base fluid. Nowadays nanofluids are also used in drug delivery systems [22] and advanced nuclear systems [23] due to enriched thermal properties. The nanofluid technology is still in its early stage and various researchers are using nanofluids as a tool to solve technological riddles of the modern society. **Figure 1** establishes big impact of small particles in view of the diverse applications of nanofluids in fields of industrial, residential, biomedical and transportation.

These days, nanoparticles are used in almost every biomedical application. Recent usage of nanotechnology in medicine and cancer therapy has attracted a lot of interest in thermal properties of nanofluid such as blood with nanoparticles suspension. Researchers have made the efforts to construct a mathematical model that shows the physical system or phenomenon nearly exact behaviour [24, 25]. Motivated by their work, we also intended to form an analytical model for the analysis of the convection currents in a horizontal nanofluid layer which is in accordance with the physical laws. Consequently, the onset of convection currents in the nanofluid layer is investigated mathematically with the help of partial differential equations. To begin with, equations are non-dimensionalized to get Rayleigh

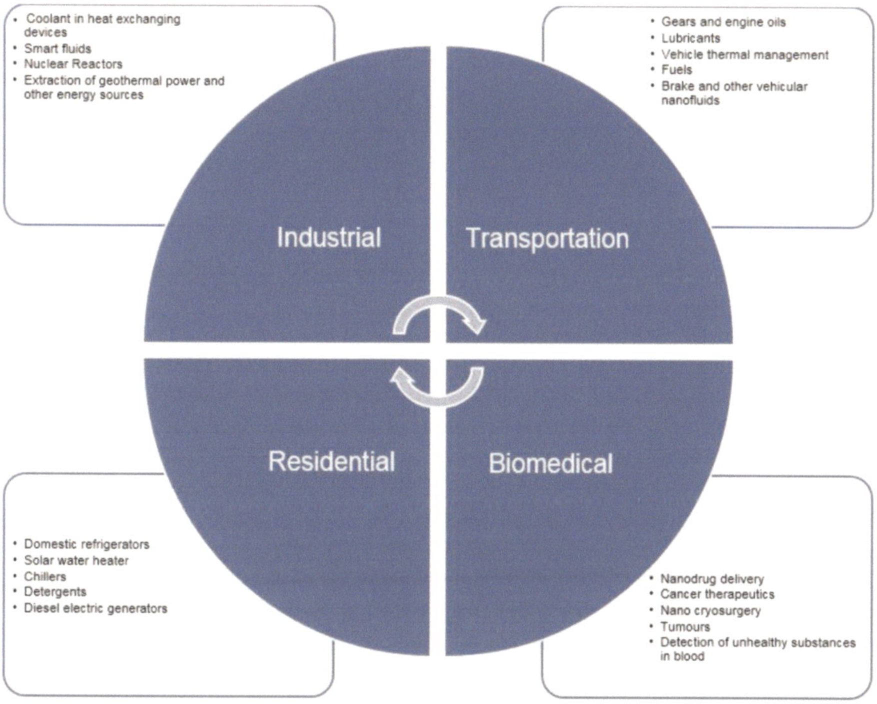


**Figure 1.**
*Applications of nanofluids.*

number in the system. Then small disturbances are added to the initial flow and new set of equations are obtained. Further PDE's are converted into ordinary differential equations using normal modes and expression for Rayleigh number is obtained. It is found that density and conductivity of nanoparticles are important parameters in deciding the stability of the system.

## 2. Instability of fluids under small temperature gradient: Rayleigh Bênard convection

The convective motions occur in a fluid layer heated underside in which a small temperature gradient is maintained across its boundaries. The maintained temperature across the boundaries must surpass a certain value before the instability can manifest itself. This Phenomenon was discovered by Bénard [26] in 1900. In most of his experiments, he found that if a fluid layer is heated underside, the layer at the bottom expands due to higher temperature. This makes the fluid density lighter at the bottom than that on the top making the system top heavy. Here viscosity and thermal diffusivity tend to oppose the convective motions but with the application of higher temperature gradient across the fluid layer, the thermal convection process gets initiated showing the pattern of cellular motions (called Bénard convection). Bénard [27] performed an experiment with metallic plate with a thin non-volatile liquid layer of 1 mm depth maintained under constant temperature.

Keeping the upper layer of fluid exposed to free air, he observed that the fluid layer was decomposed into number of cells (showing cellular motion) called Bénard cells. Thus in the standard Bénard problem, density difference due to variation in

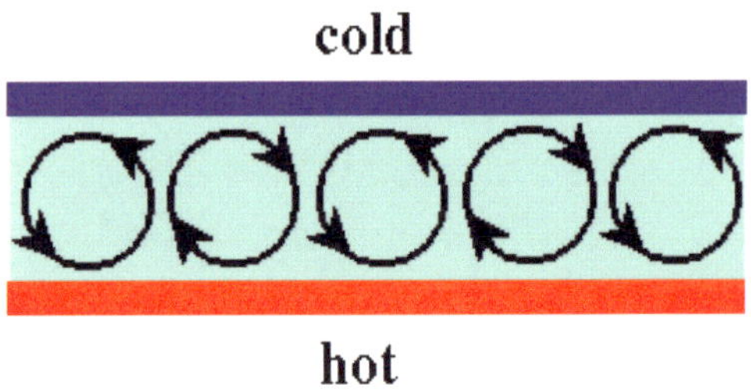


**Figure 2.**
*A view of Rayleigh-Bénard convection.*

temperature across the upper and lower boundaries of the fluid becomes the main reason for the occurrence of instability. **Figure 2** shows the schematic representation of Rayleigh-Bénard convection. Rayleigh [28] was the first person who gave an analytical treatment of the problem related to identifying the conditions responsible for breakdown of basic state. As a subsequent work carried out by Rayleigh and Bénard, thermal instability of fluids is known as Rayleigh Bénard convection. The condition for convective motions (depends on temperature gradient) can be represented in dimensionless form by the critical Rayleigh number. He figured out the condition for the instability of free surfaces by showing that the instability would occur on a large temperature gradient $\beta = -dT/dz$ in such a way that the Rayleigh number; $R_A = \alpha\beta \mathbf{g} d^4/\kappa\nu$, exceeds a certain critical value; where acceleration due to gravity is represented by g, coefficient of thermal expansion by $\alpha$, the depth of the layer by d, thermal diffusivity by $\kappa$ and kinematic viscosity is given by $\nu$. For the stabilizing viscous force, $R_A$ parameter gives the force of destabilizing buoyancy. Chandra [29] found discrepancy between the theoretical and experimental work for the convective motions in fluids when heated underside. He explained it by conducting an experiment on the layer in air and observed that instability of the fluid layer was dependent on its depth. A simplification in the partial differential equations describing the flow of compressible fluid is done by Spiegel and Veronis [30] by assuming very small depth of the layer as compared to the height. The basic equations of a fluid layer in porous medium (when heated underside) were formulated and derived by Joseph [31] by using Boussinesq approximation. The problem of thermal convection of a fluid layer has been put forward by Chandrasekhar [32] by considering the implications of various aspects of hydrodynamics and hydromagnetics. He depicted the result that addition of rotation and magnetic field increases the stability of the system. Kim et al. [33] considered the same problem of thermal convection for nanofluids. They showed that convective motion directly depends on the two physical properties (heat capacity and density) of nanoparticles and adversely depends on the conductivity of nanoparticles. Buongiorno [6] was the first scientist who formulated the conservation equations of nanofluids by assimilating the effects of diffusion due to Brownian motion and thermophoresis of nanoparticles. During his analysis, he concluded that Brownian and thermophoretic diffusion play a significant role in the absence of turbulent effects as compared to other seven mechanisms. Hwang et al. [34] treated this problem analytically and put forth the result of thermal instability of water based nanofluid with alumina nanoparticles in a rectangular container which is heated from below. They found that stability of the base fluid is enhanced by adding alumina nanoparticles and further it is enhanced by increasing the volume fraction of nanoparticles, the average temperature of the nanofluids and by decreasing the size of nanoparticles. They observed the decrease in heat transfer coefficient of nanofluids with the increase of the size of nanoparticles and decrease in the temperature of nanofluids. Tzou [35, 36] investigated the onset of convective instability of nanofluids using Buongiorno's model analytically and established that nanofluids exhibit much lower stability than regular fluids. The results depict the inverse relationship of density and heat capacity of nanoparticles with their thermal conductivity and the shape factor. The results also include that the heat transfer coefficient of nanofluid is enhanced relative to volume fraction of nanoparticles. Nield and Kuznetsov [37] reconsidered the instability problem for nanofluids to get the expression for Rayleigh number and found conditions for the existence of oscillatory convection. It was established that the buoyancy coupled with the conservation of nanoparticles lead to higher instability of nanofluids. Alloui et al. [38] considered the shallow cavity to study Rayleigh Bénard instability for nanofluids. They concluded that rate of heat transfer in nanofluids depends on the strength of convection and volume fraction of nanoparticles while the presence of nanoparticles increase the stability of the system. Thermal instability for a horizontal nanofluid layer was considered by Yadav et al. [39]. They found an expression of thermal Rayleigh number and observed that the temperature gradient delays the convective motions while volumetric fraction of nanoparticles and the ratio of the density of nanoparticles to that of base fluid have destabilizing impact on the layer. The joint behaviour of nano-effects (Brownian motion and thermophoresis) creates destabilizing effect and can reduce the values of critical Rayleigh number as compared to that of regular fluids.

## 3. Conservation equations for a nanofluid layer

We start this section with the description of Boussinesq approximation which is used to write the conservation equations of nanofluids in simplified form.

As is the case of regular fluid [32], equations of nanofluids are difficult to solve because of their non-linear character. Therefore some mathematical approximations are to be used to simplify the basic equations without violating the physical laws. The contribution of Boussinesq [40] in the solution of thermal instability problems is in the form of approximations which is after his name. This approximation has been used by a many researchers for solving different problems of fluids. Boussinesq suggested that inertial effects of density variations can be neglected as compared to its gravitational effects as such situations exist in the domain of meteorology and oceanography. So, density is assumed to be constant everywhere in the equations of motion except in the term with external force. Therefore, we change $\rho_0[1 + \alpha(T_0 - T)]$ by $\rho_0$ everywhere in the equations of motion except the term representing the external body force.

Anoop et al. [41] explained various experimental techniques using which nanoparticles can be suspended in the base fluid and that suspension remain stable for several weeks. Buongiorno [6] adopted the formalism of Bird et al. [42] and Chandrasekhar [32] to write conservation equations for nanofluids by considering nanoscale effects; Brownian diffusion and thermophoresis. A model for convective transport in regular fluids was reformulated for nanofluids to accommodate these nanoscale effects as follows.

The random motion of nanoparticles is called Brownian motion and results into the continuous collisions with the base fluid molecules. The Brownian diffusion coefficient due to Brownian motion is given by

$$D_B = \frac{k_B T}{3\pi\mu d_p}, \tag{1}$$

where $d_p$ is the nanoparticle's diameter, $k_B$ is the Boltzmann's constant and $\mu$ is the viscosity of the fluid. The nanoparticles mass flux due to Brownian diffusion, $\boldsymbol{j_{p,B}}$ is given as

$$\boldsymbol{j_{p,B}} = -\rho_p D_B \nabla \phi, \tag{2}$$

where $\phi$ is the nanoparticle volume fraction and $\rho_p$ is the nanoparticle mass density.

Thermophoresis is the phenomenon in which particles diffuse due to temperature gradient and the effect is similar to one of well-known effects of solute; Soret effect. The thermophoretic velocity is defined as.

$$\boldsymbol{V}_T = -\tilde{\beta}\frac{\mu}{\rho}\frac{\nabla T}{T} \text{ where } \tilde{\beta} = 0.26\frac{k}{2k + k_p}. \tag{3}$$

Here, $\rho$ is the overall density of the nanofluid, $k$ and $k_p$ are the thermal conductivities of the fluid and the particle material, respectively. The negative sign in thermophoretic velocity represents movement of particles down the temperature gradient (from hot to cold). The nanoparticle mass flux due to thermophoresis, $j_{p,T}$ is given as.

$$\boldsymbol{j_{p,T}} = -\rho_p \phi \boldsymbol{V}_T = -\rho_p D_T \frac{\nabla T}{T} \text{ with } D_T = \tilde{\beta}\frac{\mu}{\rho}\phi, \tag{4}$$

where $D_T$ represents the thermophoretic diffusion coefficient.

The nanoparticles mass flux due to Brownian diffusion (Eq. (1)) and thermophoresis (Eq. (4)) are used to develop a two-component model for convective transport in nanofluids with the following assumptions:

- The nanofluid flow is incompressible.
- There are no chemical reactions in the fluid layer.
- The external forces are negligible.
- The mixture is dilute with nanoparticle volume fraction less than 1%.
- The viscous dissipation is negligible in the fluid.
- The radiative heat transfer is negligible.
- The nanoparticles and base fluid are locally in thermal equilibrium.

The seven equations based on basic conservation laws with the above mentioned assumptions are given as follows.

- Equation of state (one).
- Equation of continuity (one).
- Equation of nanoparticles (one).
- Equations of motion (three).
- Equation of energy (one).

### 3.1 Equation of state

Variables of state depend only upon the state of a system. The physical quantities: $p$; the pressure, $T$; the temperature and $\rho$; the density are the variables of state. We have three thermodynamic variables and a relation between them is given as.

$$F(p, \rho, T) = 0, \tag{5}$$

For substances with which we shall be principally concerned, the equation of state can be written as

$$\rho = \rho_0[1 + \alpha(T_0 - T)], \tag{6}$$

where $T_0$ is the temperature at which $\rho = \rho_0$.

### 3.2 Equation of continuity-conservation of mass

The equation of continuity for nanofluids is

$$\frac{\partial \rho}{\partial t} + u_j \frac{\partial \rho}{\partial x_j} = -\rho \frac{\partial u_j}{\partial x_j}, \tag{7}$$

where $u_j$ is the jth component of nanofluid's velocity.
For an incompressible flow (using equation of state)

$$\frac{\partial \rho}{\partial t} + u_j \frac{\partial \rho}{\partial x_j} = 0, \tag{8}$$

so that the Eq. (7) reduces to

$$\frac{\partial u_j}{\partial x_j} = 0, \tag{9}$$

and in vector form continuity equation for nanofluid is expressed as

$$\nabla . \boldsymbol{v} = 0. \tag{10}$$

### 3.3 Equation of nanoparticles-conservation of mass

The conservation equation for nanoparticles in absence of chemical reactions is

$$\frac{\partial \phi}{\partial t} + \boldsymbol{v}.\nabla \phi = -\frac{1}{\rho_p} \nabla . \boldsymbol{j_p}, \tag{11}$$

where $t$ is the time, $\phi$ is the nanoparticles volume fraction and $\boldsymbol{j}_p$ is the diffusion mass flux for nanoparticles and as external forces are negligible $\boldsymbol{j}_p$, the sum of two diffusion terms (Brownian diffusion and thermophoresis) using Eqs. (1) and (4) can be written as

$$\boldsymbol{j_p} = \boldsymbol{j_{p,B}} + \boldsymbol{j_{p,T}} = -\rho_p D_B \nabla \phi - \rho_p D_T \frac{\nabla T}{T}, \tag{12}$$

Combining Eqs. (11) and (12), nanoparticles conservation equation becomes

$$\frac{\partial \phi}{\partial t} + \boldsymbol{v}.\nabla\phi = \nabla.\left[D_B \nabla\phi + D_T \frac{\nabla T}{T}\right]. \tag{13}$$

Eq. (13) reveals that the nanoparticles move consistently with fluid (second term of left-hand side) and possess velocity relative to fluid (right-hand side) due to Brownian diffusion and thermophoresis.

### 3.4 Equations of motion-conservation of momentum

The equation of motion is derived from Newton's second law of motion which states that

Rate of change of linear momentum = Total force.

The momentum equation for nanofluid with negligible external forces is

$$\rho_0\left(\frac{\partial \boldsymbol{v}}{\partial t} + \boldsymbol{v}.\nabla\boldsymbol{v}\right) = -\nabla p + \mu\nabla^2\boldsymbol{v} + \rho\boldsymbol{g}, \tag{14}$$

where $\rho_0$ is the nanofluid density at the reference temperature $T_0$ and the overall density of nanofluid; written as

$$\rho = \phi\rho_p + (1-\phi)\rho_f \cong \phi\rho_p + (1-\phi)\{\rho_0(1-\alpha(T-T_0))\} \tag{15}$$

Thus Eq. (14) becomes

$$\rho_0\left(\frac{\partial \boldsymbol{v}}{\partial t} + \boldsymbol{v}.\nabla\boldsymbol{v}\right) = -\nabla p + \mu\nabla^2\boldsymbol{v} + \left(\phi\rho_p + (1-\phi)\{\rho_0(1-\alpha(T-T_0))\}\right)\boldsymbol{g}. \tag{16}$$

Note that in the absence of nanoparticles, Eq. (16) reduces to momentum equation for regular fluid.

### 3.5 Equation of energy-conservation of energy

The thermal energy equation for nanofluid with the assumptions (i)–(v) is

$$(\rho c)\left[\frac{\partial T}{\partial t} + \boldsymbol{v}.\nabla T\right] = -\nabla.\boldsymbol{q} + h_p\nabla.\boldsymbol{j_p}, \tag{17}$$

where $c$ and $h_p$ are the specific heat of fluid (at constant pressure) and the specific enthalpy of nanoparticles, respectively and $\boldsymbol{q}$ is the energy flux, neglecting radiative heat transfer, the sum of heat fluxes due to conduction and nanoparticle diffusion, written as

$$\boldsymbol{q} = -k\nabla T + h_p\boldsymbol{j_p}, \tag{18}$$

Substituting Eq. (18) in Eq. (17), we get

$$(\rho c)\left[\frac{\partial T}{\partial t} + \boldsymbol{v}.\nabla T\right] = \nabla.(k\nabla T) - c_p\boldsymbol{j_p}.\nabla T. \tag{19}$$

with assumption of negligible external forces $\nabla h_p = c_p\nabla T$. Substituting Eq. (12) in Eq. (19); gives final form of thermal energy equation as

$$(\rho c)\left[\frac{\partial T}{\partial t}+\boldsymbol{v}.\nabla T\right]=k\nabla^2 T+(\rho c)_p\left[D_B\nabla\phi.\nabla T+D_T\frac{\nabla T.\nabla T}{T}\right]. \tag{20}$$

Note that if $\boldsymbol{j}_p$ is zero, Eq.(19) and hence Eq. (20) reduces to the familiar energy equation for regular fluid and therefore last two terms on right-hand side truly account for contributions of nanoparticle motion relative to fluid. Eq. (20) establishes that the transport of heat in nanofluids is possible by convection (second term on left-hand side), by conduction (first term on right-hand side), and also by virtue of nanoparticle diffusion (second and third terms on right-hand side).

Thus, Eqs. (10), (13), (16), (20) constitute the convective transport model for nanofluids which further can be solved for different parameters once the initial and boundary conditions are known. It is interesting to note that all the equations are strongly coupled meaning thereby that the one parameter depends on various other parameters.

Let us introduce non-dimensional variables to get the expression for thermal Rayleigh number as.

$$(x',y',z')=\frac{(x,y,z)}{d},t'=\frac{t\alpha_f}{d^2},\boldsymbol{v}'=\frac{\boldsymbol{v}\,d}{\alpha_f},p'=\frac{pd^2}{\mu\alpha_f},\phi'=\frac{\phi}{\phi_b},T'=\frac{T-T_0}{T_1-T_0}, \tag{21}$$

where $\alpha_f=\frac{k}{\rho C}$.

Using Eqs. (21), (10), (13), (16), (20) after dropping the dashes are

$$\nabla.\boldsymbol{v}=0, \tag{22}$$

$$\frac{\rho\alpha_f}{\mu}\left(\frac{\partial\boldsymbol{v}}{\partial t}+\boldsymbol{v}.\nabla\boldsymbol{v}\right)=-\nabla p+\nabla^2\boldsymbol{v}-\frac{\rho\boldsymbol{g}d^3}{\mu\alpha_f}\hat{k}+R_A T\hat{k}-\frac{\left(\rho_p-\rho\right)\phi_b\boldsymbol{g}\,d^3}{\mu\alpha_f}\phi\,\hat{k}, \tag{23}$$

$$\frac{\partial T}{\partial t}+\boldsymbol{v}.\nabla T=\nabla^2 T+\frac{(\rho C)_P}{\rho C}\phi_b\frac{D_B}{\alpha_f}\nabla\phi.\nabla T+\frac{D_T(T_1-T_0)}{D_B T_0\phi_b}\frac{(\rho C)_P}{\rho C}\phi_b\frac{D_B}{\alpha_f}\nabla T.\nabla T, \tag{24}$$

$$\frac{\partial\phi}{\partial t}+\boldsymbol{v}.\nabla\phi=\frac{D_B}{\alpha_f}\nabla^2\phi+\frac{D_T(T_1-T_0)}{D_B T_0\phi_b}\frac{D_B}{\alpha_f}\nabla^2 T, \tag{25}$$

where thermal Rayleigh number $R_A=\frac{\rho\,g\,\beta_T d^3(T_1-T_0)}{\mu\alpha_f}$.

## 4. Initial and perturbed flow

At the initial state, it is assumed that nanoparticle volume fraction is constant and fluid layer is still while temperature and pressure vary in horizontal direction. We get initial solution of Eqs. (22)–(25) using the fact that thermal diffusivity is very large as compared to Brownian diffusion coefficient (refer Buongiorno [1]) as

$$v_i=0,\phi_i=1,T_i=1-z \tag{26}$$

Let us add perturbations to initial solution and write

$$(\boldsymbol{v},p,T,\phi)=\left(\boldsymbol{v}_i+\tilde{\boldsymbol{v}},p_i+\tilde{p},T_i+\tilde{T},\ \phi_i+\tilde{\phi}\right). \tag{27}$$

The Eq. (27) in Eqs. (22)–(25) give

$$\nabla.\tilde{\boldsymbol{v}} = 0, \tag{28}$$

$$\frac{\rho\alpha_f}{\mu}\frac{\partial\tilde{\boldsymbol{v}}}{\partial t} = -\nabla\tilde{p} + \nabla^2\tilde{\boldsymbol{v}} + R_A\tilde{T}\hat{k} - \frac{\left(\rho_p - \rho\right)\phi_b \boldsymbol{g}\, d^3}{\mu\alpha_f}\tilde{\phi}\hat{k}, \tag{29}$$

$$\frac{\partial\tilde{T}}{\partial t} - \tilde{u}_3 = \nabla^2\tilde{T} - \frac{(\rho C)_P}{\rho C}\phi_b\frac{D_B}{\alpha_f}\frac{\partial\tilde{\phi}}{\partial z} - 2\frac{D_T(T_1 - T_0)}{D_B T_0\phi_b}\frac{(\rho C)_P}{\rho C}\phi_b\frac{D_B}{\alpha_f}\frac{\partial\tilde{T}}{\partial z}, \tag{30}$$

$$\frac{\partial\tilde{\phi}}{\partial t} = \frac{D_B}{\alpha_f}\nabla^2\tilde{\phi} + \frac{D_T(T_1 - T_0)}{D_B T_0\phi_b}\frac{D_B}{\alpha_f}\nabla^2\tilde{T}. \tag{31}$$

Making use of the identity *curlcurl* ≡ *graddiv* $-\ \nabla^2$ curlcurl = graddiv $-\ \nabla^2$ on Eq. (29) together with Eq. (28), we get

$$\frac{\rho\alpha_f}{\mu}\frac{\partial}{\partial t}\left(\nabla^2\tilde{u}_3\right) - \nabla^4\tilde{u}_3 = R_A\nabla_H^2\tilde{T} - \frac{\left(\rho_p - \rho\right)\phi_b \boldsymbol{g}\, d^3}{\mu\alpha_f}\nabla_H^2\tilde{\phi}, \tag{32}$$

where $\nabla_H^2 = \frac{\partial^2}{\partial x^2} + \frac{\partial^2}{\partial y^2}.$.

## 5. Method of normal modes

To change PDE's to ODE's, Eqs. (30)–(32) are solved using normal mode analysis and perturbed variables are written as

$$\left(\tilde{u}_3, \tilde{T}, \tilde{\phi}\right) = (W(z), \mathrm{T}(z), \Phi(z))\exp\left(ik_x x + ik_y y + st\right), \tag{33}$$

Thus above mentioned equations reduce to

$$\left(\left(D^2 - \alpha^2\right)^2 - \frac{s\rho\alpha_f}{\mu}\left(D^2 - \alpha^2\right)\right)W - R_A\alpha^2\mathrm{T} + \frac{\left(\rho_p - \rho\right)\phi_b \boldsymbol{g}\, d^3}{\mu\alpha_f}\alpha^2\Phi = 0, \tag{34}$$

$$W + \left(\left(D^2 - \alpha^2\right) - s - 2\frac{D_T(T_1 - T_0)}{D_B T_0\phi_b}\frac{(\rho C)_P}{\rho C}\phi_b\frac{D_B}{\alpha_f}D\right)\mathrm{T} - \frac{(\rho C)_P}{\rho C}\phi_b\frac{D_B}{\alpha_f}D\Phi = 0, \tag{35}$$

$$\left(\frac{D_B}{\alpha_f}\left(D^2 - \alpha^2\right) - s\right)\Phi + \frac{D_T(T_1 - T_0)}{D_B T_0\phi_b}\frac{D_B}{\alpha_f}\left(D^2 - \alpha^2\right)\mathrm{T} = 0, \tag{36}$$

where $D \equiv \frac{d}{dz}$, $\alpha = \left({k_x}^2 + {k_y}^2\right)^{1/2}$. Using one term Galerkin weighted residual method and free-free boundaries conditions

$$W = D^2W = \mathrm{T} = 0 \text{ at } z = 0 \text{ and } z = 1. \tag{37}$$

We write

$$W = A\sin\pi z, \text{ and } T = B\sin\pi z, \tag{38}$$

using of the orthogonality to the functions; gives eigenvalue equation as

$$\left((J+s)\left(\frac{JD_B}{\alpha_f}+s\right)\left(J^2+\frac{\rho\alpha_f Js}{\mu}\right)-\alpha^2\left(\frac{JD_T\left(\rho_p-\rho\right)(T_1-T_0)\mathbf{g}\,d^3}{D_B\mu\alpha_f^2T_0}+R_A\left(\frac{JD_B}{\alpha_f}+s\right)\right)\right)$$
$$-\alpha^2J\left(\frac{JD_T\left(\rho_p-\rho\right)(T_1-T_0)\mathbf{g}\,d^3}{D_B\mu\alpha_f^2T_0}+R_A\left(\frac{JD_T\left(\rho_p-\rho\right)(T_1-T_0)\mathbf{g}\,d^3}{D_B\mu\alpha_f^2T_0}+s\right)\right)=0 \tag{39}$$

where $J=\pi^2+\alpha^2$.

## 6. Results and discussion

### 6.1 Stationary convection

For non-oscillatory motions $s=0$, this gives the expression for $R_A$ from Eq. (39) as

$$R_A=\frac{J^3}{\alpha^2}-\frac{D_T\left(\rho_p-\rho\right)(T_1-T_0)\mathbf{g}\,d^3}{D_B\mu\alpha_fT_0}. \tag{40}$$

where $D_B=\frac{k_BT}{3\pi\mu d_p}$ and $D_T=\tilde{\beta}\frac{\mu}{\rho}\phi$ with $\tilde{\beta}=0.26\frac{k}{2k+k_p}$ as given by Nield and Kuznetsov [35].

Also

$$R_A=\frac{J^3}{\alpha^2}-\frac{\left(\rho_p-\rho\right)}{2k+k_p}A; \tag{41}$$

where $A$ depends on the base fluid properties.

## 7. Discussions on analytical results using various metallic/non-metallic nanoparticles

**Table 2** shows the ratios of density to conductivity of various metallic/non-metallic nanoparticles and density 997.1 and conductivity 0.613 of water is used.

It is observed that ratio of density to conductivity is accountable for hastening the onset of convection in the system. The ratio is more for non-metals than metals establishes the lesser stability of non-metallic nanoparticles than metals. Alumina is

| Physical properties | Al | Cu | Ag | Fe | $Al_2O_3$ | $SiO_2$ | CuO | $TiO_2$ |
|---|---|---|---|---|---|---|---|---|
| $\rho\,(\text{Kg/m}^3)$ | 2700 | 9000 | 10,500 | 7900 | 3970 | 2600 | 6510 | 4250 |
| $\text{k}\,(\text{W/m}K)$ | 237 | 401 | 429 | 80 | 40 | 10.4 | 18 | 8.9 |
| $\rho_p/\text{k}_p$ | 11.3 | 22.4 | 24.47 | 98.7 | 99.25 | 250 | 361.6 | 477.5 |

**Table 2.**
*Ratios of density to conductivity of metallic and non-metallic nanoparticles.*

most stable and titanium oxide is least stable among the nanoparticles under consideration. Density is found to be more influential than conductivity towards deciding the onset of convection in the layer.

## 8. Conclusions

Tremendous applications of nanofluids in pharmaceutical industry with respect to drug invention and cancer imaging motivated the scientists to study convection currents in fluid layer mathematically as well as experimentally. In the present work, the onset of instability in layer is studied under small temperature difference with the help of equations based on conservation laws. The expression of non-dimensional number Rayleigh number is found analytically which decides the instability of the system. Approximations are made whenever needed without violating the necessary physics to get the useful results. Analysis reveals that lesser the ratio of density to conductivity, higher is the stability of the layer. It is found that convection currents majorly depends on density and conductivity and precisely concluding density is more pronounced property than conductivity of nanoparticles. Metallic oxides make the system more stable than metallic nanoparticles in the fluid.

**Author details**

Jyoti Sharma[1*] and Urvashi Gupta[2]

1 U.I.E.T, Panjab University, Chandigarh, India

2 Dr. S. S. Bhatnagar University Institute of Chemical Engineering and Technology, Panjab University, Chandigarh, India

*Address all correspondence to: jyoti.maths@gmail.com

## References

[1] Maxwell JC. Electricity and Magnetism. Oxford, UK: Clarendon Press; 1873

[2] Choi S. Enhancing thermal conductivity of fluids with nanoparticles. In: Siginer DA, Wang HP, editors. Development and Applications of Non-Newtonian Flows. FED- 231/ MD-Vol. 66. New York; The American Society of Mechanical Engineers (ASME); 1995. pp. 99-105

[3] Masuda H, Ebata A, Teramae K, Hishinuma N. Alteration of thermal conductivity and viscosity of liquid by dispersing ultra fine particles. Netsu Bussei. 1993;**7**:227-233

[4] Eastman JA, Choi SUS, Li S, Yu W, Thompson LJ. Anomalously increased effective thermal conductivities of ethylene glycol-based nanofluids containing copper nanoparticles. Applied Physics Letters. 2001;**78**:718-720

[5] Das SK, Putra N, Thiesen P, Roetzel W. Temperature dependence of thermal conductivity enhancement for nanofluids. ASME Journal of Heat Transfer. 2003;**125**:567-574

[6] Buongiorno J. Convective transport in nanofluids. ASME Journal of Heat Transfer. 2006;**128**(3):240-250

[7] Choi SUS, Zhang ZG, Yu W, Lockwood FE, Grulke EA. Anomalous thermal conductivity enhancement in nanotube suspensions. Applied Physics Letters. 2001;**79**:2252-2254

[8] Wang X, Xu X, Choi SUS. Thermal conductivity of nanoparticle fluid mixture. Journal of Thermophysics and Heat Transfer. 1999;**13**(4):474-480

[9] Murshed SMS, Leong KC, Yang C. Enhanced thermal conductivity of $TiO_2$-water based nanofluids. International Journal of Thermal Sciences. 2005;**44**: 367-373

[10] Choi SUS. Nanofluids: From vision to reality through research. Journal of Heat Transfer. 2009;**131**(3):033106

[11] Wen D, Lin G, Vafaei S, Zhang K. Review of nanofluids for heat transfer applications. Particuology. 2009;**7**: 141-150

[12] Das SK, Choi SUS. A review of heat transfer in nanofluids. Advances in Heat Transfer. 2009;**41**:81-197

[13] Ding Y, Alias H, Wen D, Williams RA. Heat transfer of aqueous suspension of carbon nanotubes (CNT nanofluids). International Journal of Heat and Mass Transfer. 2006;**49**: 240-250

[14] Das SK, Choi SUS, Yu W, Pradeep T. Nanofluids: Science and Technology. Hoboken, New York: Wiley; 2008

[15] Chen G. Ballistic-diffusive heat conduction equations. Physical Review Letters. 2001;**86**:2297-2300

[16] Syam Sundar L, Sharma KV. Thermal conductivity enhancement of nanoparticles in distilled water. International Journal of Nanoparticles. 2008;**1**(1):66-77

[17] Xuan Y, Li Q. Investigation on convective heat transfer and flow features of nanofluids. Journal of Heat Transfer. 2003;**125**:151-155

[18] Yu W, Choi SUS. The role of interfacial layers in the enhanced thermal of nanofluids: A renovated Maxwell model. Journal of Nanoparticle Research. 2003;**5**(1–2):167-171

[19] Vassallo P, Kumar R, Amico SD. Pool boiling heat transfer experiments in silica-water nano-fluids. International Journal of Heat and Mass Transfer. 2004;**47**:407-411

[20] Chon CH, Kihm KD. Thermal conductivity enhancement of nanofluids by Brownian motion. ASME Journal of Heat Transfer. 2005;**127**:810

[21] Pak BC, Cho Y. Hydrodynamic and heat transfer study of dispersed fluids with submicron metallic oxide particles. Journal of Experimental Heat Transfer. 1998;**11**(2):151-170

[22] Kleinstreuer C, Li J, Koo J. Microfluidics of nano-drug delivery. International Journal of Heat and Mass Transfer. 2008;**51**:5590-5597

[23] Buongiorno J, Hu W. Nanofluid coolants for advanced nuclear power plants. In: Proceedings of ICAPP'05, Seoul; 2005. Paper No. 5705

[24] Saedi Ardahaie S, Jafarian Amiri A, Amouei A, Hosseinzadeh K, Ganji DD. Investigating the effect of adding nanoparticles to the blood flow in presence of magnetic field in a porous blood arterial. Informatics in Medicine Unlocked. 2018;**10**:71-81. ISSN: 2352-9148

[25] Sayed HM, Aly EH, Vajravelu K. Influence of slip and convective boundary conditions on peristaltic transport of non-Newtonian nanofluids in an inclined asymmetric channel. Alexandria Engineering Journal. 2016; **55**(3):2209-2220. ISSN: 1110-0168

[26] Bénard H. Les Tourbillons cellulaires dans une nappe liquide. Revue Générale des Sciences Pures et Appliquées. 1900;**11**:1261-1271; 1309-1328

[27] Bénard H. Les tourbillions cellularies dans une nappe liquide transportant de la chaleur par convection en regime permanent. Annales de Chimie et de Physique. 1901;**23**:62-144

[28] Rayleigh L. On convective currents in a horizontal layer of fluid when the higher temperature is on the under side. Philosophical Magazine. 1916;**32**: 529-546

[29] Chandra K. Instability of fluids heated from below. Proceedings. Royal Society of London. 1938;**A164**:231-242

[30] Spiegal EA, Veronis G. On the Bousinessq approximation for a compressible fluid. The Astrophysical Journal. 1960;**131**:442

[31] Joseph DD. Stability of Fluid Motions. Vol. 1 & 2. Berlin: Springer-Verlag; 1976

[32] Chandrasekhar S. Hydrodynamic and Hydromagnetic Stability. New York: Dover Publication; 1981

[33] Kim J, Kang YT, Choi CK. Analysis of convective instability and heat transfer characteristics of nanofluids. Physics of Fluids. 2004;**16**:2395-2401

[34] Hwang KS, Lee JH, Jang SP. Buoyancy driven heat transfer of water based $Al_2O_3$ nanofluids in a rectangular cavity. International Journal of Heat and Mass Transfer. 2007;**50**:4003-4010

[35] Tzou DY. Instability of nanofluids in natural convection. ASME Journal of Heat Transfer. 2008;**130**:372-401

[36] Tzou DY. Thermal instability of nanofluids in natural convection. International Journal of Heat and Mass Transfer. 2008;**51**:2967-2979

[37] Nield DA, Kuznetsov AV. The onset of convection in a horizontal nanofluid layer of finite depth. European Journal of Mechanics - B/Fluids. 2010;**29**: 217-223

[38] Alloui Z, Vasseur P, Reggio M. Natural convection of nanofluids in a shallow cavity heated from below. International Journal of Thermal Sciences. 2010;**50**(3):385-393

[39] Yadav D, Agrawal GS, Bhargava R. Rayleigh Bénard convection in

nanofluid. International Journal of Applied Mathematics and Mechanics. 2011;7(2):61-76

[40] Boussinesq J. Théorie Analytiquede la Chaleur. Vol. 2. Paris: Gauthier-Villars; 1903. p. 172

[41] Anoop KB, Das SK, Sundararajan T. Effect of particle size on the convective heat transfer in nanofluid in the developing region. International Journal of Heat and Mass Transfer. 2009;**52**: 2189-2195

[42] Byron Bird R, Stewart WE, Lightfoot EN. Transport Phenomena. New York: Wiley; 1960